# Measuring Maturity in Complex Engineering Projects

João Carlos Araújo da Silva Neto
*Magnesita SA, Brazil*

Ítalo Coutinho
*Saletto Engenharia de Serviços, Brazil*

Gustavo Teixeira
*PM BASIS, Brazil*

Alexandro Avila de Moura
*Paranapanema SA, Brazil*

A volume in the Advances in Civil and Industrial Engineering (ACIE) Book Series

IGI Global
DISSEMINATOR OF KNOWLEDGE

Published in the United States of America by
IGI Global
Engineering Science Reference (an imprint of IGI Global)
701 E. Chocolate Avenue
Hershey PA, USA 17033
Tel: 717-533-8845
Fax: 717-533-8661
E-mail: cust@igi-global.com
Web site: http://www.igi-global.com



Library of Congress Cataloging-in-Publication Data

Names: Silva Neto, Joao Carlos Araujo da, 1976- author. | Coutinho, Italo, 1976- author. | Teixeira, Gustavo, 1976- author. | Moura, Alexandro Avila de, author.
Title: Measuring maturity in complex engineering projects / by Joao Carlos Araujo da Silva Neto, Italo Coutinho, Gustavo Teixeira, and Alexandro Avila de Moura.
Description: Hershey, PA : Engineering Science Reference, [2018] | Includes bibliographical references.
Identifiers: LCCN 2017056615| ISBN 9781522558644 (hardcover) | ISBN 9781522558651 (ebook)
Subjects: LCSH: Project management--Evaluation. | Construction projects--Management--Case studies. | Mining engineering--Management--Case studies. | Electric power-plants--Management--Case studies.
Classification: LCC TA190 .S587 2018 | DDC 620.0068/4--dc23 LC record available at https://lccn.loc.gov/2017056615

This book is published in the IGI Global book series Advances in Civil and Industrial Engineering (ACIE) (ISSN: 2326-6139; eISSN: 2326-6155)

British Cataloguing in Publication Data
A Cataloguing in Publication record for this book is available from the British Library.

All work contributed to this book is new, previously-unpublished material.
The views expressed in this book are those of the authors, but not necessarily of the publisher.

For electronic access to this publication, please contact: eresources@igi-global.com.

# Advances in Civil and Industrial Engineering (ACIE) Book Series

ISSN:2326-6139
EISSN:2326-6155

Editor-in-Chief: Ioan Constantin Dima, University Valahia of Târgovişte, Romania

**MISSION**

Private and public sector infrastructures begin to age, or require change in the face of developing technologies, the fields of civil and industrial engineering have become increasingly important as a method to mitigate and manage these changes. As governments and the public at large begin to grapple with climate change and growing populations, civil engineering has become more interdisciplinary and the need for publications that discuss the rapid changes and advancements in the field have become more in-demand. Additionally, private corporations and companies are facing similar changes and challenges, with the pressure for new and innovative methods being placed on those involved in industrial engineering.

The **Advances in Civil and Industrial Engineering (ACIE) Book Series** aims to present research and methodology that will provide solutions and discussions to meet such needs. The latest methodologies, applications, tools, and analysis will be published through the books included in **ACIE** in order to keep the available research in civil and industrial engineering as current and timely as possible.

**COVERAGE**

- Ergonomics
- Productivity
- Operations research
- Production Planning and Control
- Transportation Engineering
- Materials Management
- Structural Engineering
- Engineering Economics
- Hydraulic Engineering
- Urban Engineering

IGI Global is currently accepting manuscripts for publication within this series. To submit a proposal for a volume in this series, please contact our Acquisition Editors at Acquisitions@igi-global.com or visit: http://www.igi-global.com/publish/.

The Advances in Civil and Industrial Engineering (ACIE) Book Series (ISSN 2326-6139) is published by IGI Global, 701 E. Chocolate Avenue, Hershey, PA 17033-1240, USA, www.igi-global.com. This series is composed of titles available for purchase individually; each title is edited to be contextually exclusive from any other title within the series. For pricing and ordering information please visit http://www.igi-global.com/book-series/advances-civil-industrial-engineering/73673. Postmaster: Send all address changes to above address. 

## Titles in this Series

*For a list of additional titles in this series, please visit:*
*https://www.igi-global.com/book-series/advances-civil-industrial-engineering/73673*

***Contemporary Strategies and Approaches in 3-D Information Modeling***
Bimal Kumar (Glasgow Caledonian Univerity, UK)
Engineering Science Reference • ©2018 • 313pp • H/C (ISBN: 9781522556251) • US $205.00

***New Approaches, Methods, and Tools in Urban E-Planning***
Carlos Nunes Silva (University of Lisbon, Portugal)
Engineering Science Reference • ©2018 • 407pp • H/C (ISBN: 9781522559993) • US $205.00

***Designing Grid Cities for Optimized Urban Development and Planning***
Guiseppe Carlone (Independent Researcher, Italy) Nicola Martinelli (Politecnico di Bari, Italy) and Francesco Rotondo (Politecnico di Bari, Italy)
Engineering Science Reference • ©2018 • 305pp • H/C (ISBN: 9781522536130) • US $210.00

***Design Solutions for nZEB Retrofit Buildings***
Elzbieta Rynska (Warsaw University of Technology, Poland) Urszula Kozminska (Warsaw University of Technology, Poland) Kinga Zinowiec-Cieplik (Warsaw University of Technology, Poland) Joanna Rucinska (Warsaw University of Technology, Poland) and Barbara Szybinska-Matusiak (Norwegian University of Science and Technology, Norway)
Engineering Science Reference • ©2018 • 362pp • H/C (ISBN: 9781522541059) • US $225.00

***Handbook of Research on Perception-Driven Approaches to Urban Assessment and Design***
Francesco Aletta (University of Sheffield, UK) and Jieling Xiao (Birmingham City University, UK)
Engineering Science Reference • ©2018 • 641pp • H/C (ISBN: 9781522536376) • US $295.00

***Convective Heat and Mass Transfer in the Free Atmosphere Emerging Research and ...***
Kaliyeva Kulyash (Lorraine University, France)
Engineering Science Reference • ©2018 • 127pp • H/C (ISBN: 9781522530503) • US $135.00

*For an entire list of titles in this series, please visit:*
*https://www.igi-global.com/book-series/advances-civil-industrial-engineering/73673*

IGI Global
DISSEMINATOR OF KNOWLEDGE

701 East Chocolate Avenue, Hershey, PA 17033, USA
Tel: 717-533-8845 x100 • Fax: 717-533-8661
E-Mail: cust@igi-global.com • www.igi-global.com

# Table of Contents

## Section 3
## A Model to Measure Complex Projects



## Section 4
## Case Studies: Maturity, Complex Projects, Engineering, and Construction

# Foreword

The competitiveness grows fast in the world. New companies have appeared at the same time that other, larges and mature in their field, are disappearing.

Good image and reputation aren't enough to survive on this current market. No company in the world lives just from its history, needs to follow the tendencies, in the right speed.

To be creative, to speed up solution, to response as quick as possible the market's needs, can mean to live or to dead.

The demand could be for a new product, or perhaps only a few changes, or scaling up, or even a new one. Sometimes to increment the downstream, or reduce the production cost.

In all of these cases, a new project will be necessary, more or less complex, but be sure, it is primordial.

Most of the companies have a project team, and for large project needs to contract an external one. These companies have guidelines, savvy projects managers and spite of that, 30 percent of large capital-intensive project costs is wasted.

Integration of different disciplines, such as, engineering, geology, environmental, licenses, infrastructure, in order to guarantee the complex information flow, and the right time for the actions that needs to be done is very important for the success of the project.

This book, by these excellent authors, has an easy language, a differentiated didactic, which one will give to the managers the needed security, with the rights tools to address the main issues, in the organized way.

On top of that will be possible to check the maturity of the project and if necessary the main drivers to get it.

With some cases and examples of complex projects the reading becomes light, pleasant and very useful for all the levels of professional.

Together with the high level of technical information this book becomes a source of consultation, adding knowledge for mature professional as well as for the new ones interested in this matter.

Congratulation for the authors, very good contribution! Readers, enjoy this book!

A lot of useful information.

*Sebastiao Costa Filho*
*ArcelorMittal Mineração, Brazil*

# Preface

## INTRODUCTION

In the world of today, a globalized world, two issues are become strongly important when we are dealing with project implementation: project management maturity and the increasing of projects complexity. These two topics are the core of this book.

## PROJECT MANAGEMENT MATURITY

Given the globalization of the market, the players increased their capability to react quickly for any market movement in order to keep the market share. In addition to this point, the companies realize that we have just one chance to try, given how fast their players react!

Failures in implementing projects can result mainly in the misalignment against the primary strategic plan, which (as the lector is familiar) would drive the company to achieve the mission and vision. Because of that, companies during the past years and current ones are struggling to improve and even to create the area of project management and to ensure the strategic alignment.

To ensure that the company is in the right way during the implementation of the project management methodology, a model to measure the maturity and to provide the drivers to the company achieve its excellence in this field is considered the best tool and the path to speed up the project management implementation.

Today in the market we have different types of maturity model that could be applied in several types of projects. The difference is the way that they are organized, but they are goods tools to start the application.

## COMPLEX PROJECTS

Additional problems appear when we try to apply the measurement maturity model over the complex projects. Complex projects have additional points not primarily measured by the existing models such as: ambiguity and multiple stakeholders.

On ambiguities, it is convenient to highlight that they are conceptually related to enterprise environmental factors and also related to the concept of risks.

Concerning stakeholders, it should be mentioned that different stakeholders may have different expectations, which can compete with each other and create conflicts within the project.

Our discussion will also consider the combination of these two characteristics, that is, projects that have stakeholders with ambiguous positions.

A third characteristic that we will consider in our discussions of complex projects is multinationality. It is understood as the characteristic of projects whose activities are dispersed in diverse places of the globe, something usual nowadays. So, it can make the issue of (multiple) stakeholders more complex by adding the ingredient of potentially meaningful cultural differences between them. These cultural differences can bring new ambiguities to the project.

### Maturity in Complex Projects

At this point, as we already present some ideas about the need of project management maturity and also about the increasing number of complex projects, we are able to establish the main problem addressed for this book.

The objective this book is discuss the best way to promote the measuring of project management maturity in complex engineering projects.

In addition to this discussion and the information shared with the lector, this book will propose models that will be adding additional important pillars to be applied over the measurement of complexity projects and investments. The lack of these pillars can be recognized as sources of failures in considerable investments in the globe.

Beyond benefits listed above, the book intents to provide insights for further researches and developments of its subject: maturity in complex projects.

## WHO CAN USE THIS BOOK?

The target audience of this book are graduation and master degree students as well as professional which are dealing with (lack of) project management maturity, complex projects or both.

Students will find in first three sections a source of important concepts and references.

Professionals will get important insights to deal with these professional challenges based on same concepts and references, but mainly based on the three case studies.

We believe that some of these professionals can find the same (or similar) problems that they are living described in case studies. If it happens, the case studies will become an important source of insights for these professionals.

## ORGANIZATION OF THE BOOK

### Book Sections

To achieve this objective, this book will provide: concepts about project management and complex projects, an embracing vision of maturity model genesis, a deep discussion about the main ones and case studies in different sectors. So, this book has been divided in four sections.

#### Section 1: Complex Projects to Engineering and Construction

In this section, the book brings to the reader the understanding and knowledge about the culture and main characteristics regarding the complex projects and how they are in fact considered complex. In this part of the book, the authors start already discussing the maturity models in order to better compare the use of the proposed technique as our proposal in this book.

#### Section 2: Theatrical Background – Understanding the Complex Projects and Maturity

Here, the book will explain what the nature of complex projects is and how organizations can deal with them. It will analyze some organizations aspects as culture, structure, knowledge management and communication, which are relevant in this context.

Since the book comes with a proposal to introduce the most common maturity models in order to propose the new one, one of the objectives of this session is to explain with more details four of the important models existing today in the community of project management:

- Capability Maturity Model (CMM);
- Maturity Model in Project Management (MMGP);
- Organizational Project Management Maturity Model (OPM3);
- Project Management Maturity Model (PMMM).

This section also discusses one of the top characteristics of the complexity in project which is related to the stakeholder management. This discussion comprises several criteria to classify the different kinds for stakeholders: internal or external, direct or indirect, and others.

Is important to say that, in the last decades a new understanding has emerged, which considers that the objective of the organizations is to take care of the interests of all the stakeholders, having no more or less legitimate interests. Add to this the understanding that any stakeholder can influence the profit generation of an organization, and thus they become of great interest to it.

## Section 3: A Model to Measure Complex Projects

In this session, the reader will have the first contact with the proposed maturity model with the objective to measure complex projects. The idea here is to explain how it was organized, the idea behind of the pillars and also the proposal for application.

## Section 4: Case Studies – Maturity, Complex Projects, Engineering, and Construction

This is considered one of the most important session of the book. Here, the idea is to demonstrate to the reader the application of the model in 03 big sources of complex engineering projects: mining, civil construction and energy. The proposal here is to confirm the application of the model over these scenarios.

## Book Chapters

The book is organized into 12 chapters and one appendix. A brief description of each of the chapters follows:

- Chapter 1, "Engineering and Construction Projects Characteristics," includes details and concepts about Engineering and Construction initiatives and the relationship with the project management.
- Chapter 2, "Project Management and Maturity in Complex Engineering Projects," shows models of maturity usual for Engineering and Construction.
- Chapter 3, "Culture, Knowledge Management, and Maturity in Complex Engineering Projects," shows how these four elements react in Engineering and Construction Projects.
- Chapter 4, "Introduction to Complex Projects," explores concepts about complexity and its use in Engineering and Construction.
- Chapter 5, "Types of Complex Engineering Projects," gives models for complex projects.
- Chapter 6, "Maturity and Models Around Project Management," explores usual models for Maturity in Project Management.
- Chapter 7, "Choice of Maturity Model to Complex Engineering Projects," explores how to choose the correct and usual Maturity Model.
- Chapter 8, "Stakeholders in Engineering and Construction Projects," examines interests and the Engineering and Construction projects.
- Chapter 9, "How to Apply the Model to Measuring Complex Engineering Projects," gives a manual to use a model to measure maturity in projects.
- Chapter 10, "Project Maturity Analysis in the Mining Sector," examines a research about Maturity in Mining Sector – a practical application.
- Chapter 11, "Project Maturity Analysis in Energy Sector," is a research about Maturity in Energy Sector – a practical application.
- Chapter 12, "Project Maturity Analysis in Civil Construction," examines a research about Maturity in Civil Construction Sector – a practical application.
- Appendix, "Resume of the Maturity Assessment in Mining, Energy, and Civil Construction Sectors," gives a template and model.

We invite you to have a good lecture of this book and use it as a source of knowledge to provide you a better application of the project management tools. This was an important contribution for the field of project management.

# Section 1
# Complex Projects to Engineering and Construction

# Chapter 1
# Engineering and Construction Projects Characteristics


## ABSTRACT

*Companies try to use better techniques to deal with complex projects. The maturity of companies to apply these better techniques can guarantee project success or not. The environment, where the engineering and construction project is developed, also has influence over the success. The maturity of a company's project management processes is also a factor of success. This chapter aims to talk about concepts related to engineering and construction initiatives.*


## CONCEPT OF ENGINEERING AND CONSTRUCTION PROJECTS

Different kinds of project (as the development of new housing units, new industrial plants, highways or railways) comprises engineering and construction phases (Coutinho, 2009). They require multidisciplinary knowledge and high level of planning.

These kinds of projects usually will have following characteristics or demands: application of accurate engineering techniques, high amounts of financial resources, large teams, time (of conclusion) pressure, several kinds of risks, including project environmental risks, and different expectations about quality.

DOI: 10.4018/978-1-5225-5864-4.ch001

Engineering and Construction projects may be unpublished, or reforms, or repowering and or maintenance of large recoveries, aimed at understanding the application of a Capital Expenditure (CAPEX) from investors or the State itself.

The engineering of a project comprises a coordinated series of activities or tasks performed by engineers and designers from one or more engineering disciplines or departments. Its tasks consist of such things as performing calculations, writing specifications, preparing bids, reviewing equipment proposals and evaluating or selecting equipment and preparing various lists, such as equipment and materials lists, and drawings such as electrical, piping and instrumentation diagrams, physical layouts and other drawings used in design and construction.

Some companies have in house staff to handle small projects, while other larger companies have a department that does internal project engineering. Large projects are typically outsourced to engineering companies. Staffing at engineering companies varies according to the work load. In these companies the employment may only last until an individual's tasks are completed.

Based on the level of uncertainty surrounding the project, we can assume that risks will be closer to the project from the initiation until its closure.

Now we begin to understand the importance, for a company, to have project management processes in place to guarantee or to decrease the level of uncertainty of a project (Kerzner, 2003). We have to understand that it's not enough only applying concepts of project management but ensuring that they are properly implemented and being followed by the respective team of the project.

At this point we can realize the importance of an instrument to evaluate if a company has maturity in its project management processes. The unique way to check the reliability of a project management process is measuring its maturity using a defined model. For this company's board, this maturity simplifies the process to invest financial resources and become them more confident about the project approval and development process.

The objective of this book is to talk about the models used to check the maturity of project management process of a company as well as to discuss the use of this model in complex projects.

## COMMON CHARACTERISTICS OF ENGINEERING AND CONSTRUCTION PROJECTS

When we look at a building site, the assembly of a soccer stadium or the construction of a hydroelectric power plant, at first we clearly perceive that these are works with different objectives, but done a deeper evaluation we can perceive some similarities (Coutinhom 2009).

The characteristics common to these projects will determine how your administration will be and what care we should take (PMI, 2017). At first, we can find the following similarities between them:

- Underestimated deadline;
- Shortage of skilled labor force;
- High turnover of labor;
- High risk of accident due to hostile environment;
- Significant numbers of people and companies involved;
- Relevant stakeholder interference;
- High impact in the site of deployment;
- Use of pioneering technology;
- Sometimes embedded in a political setting.

Each industry suffers with the impact of the aforementioned factors, but it is undeniable that they are always present in the engineering and construction environments.

Understanding each one of them also means determining the execution strategy of the venture, also known as the Attack Plan. The sizing of communication resources and methods also depends on the type of project and its particularities.

### Underestimated Deadline

We can find several reasons to explain, even to justify, deadlines that are not consistent with the realization of the enterprise: competition, strategic need of the company, pressure from public agencies, interest and pressure from society, and incorrectly estimation between the parties.

Regardless of the reason, in practice what we see are projects with ever shorter deadlines, but not necessarily being executed as planned (Dinsmore, 2004). Increasingly recurrent delays, an increase in the number of accidents, staff with health problems due to the stress of the work environment, these are some of the consequences of the underestimated deadlines.

## Shortage of Skilled Labor Force

Even in an informal chat, we can easily realize the shortage of labor, especially in the construction and industrial assembly environment.

One of the reasons given for a shortage of labor is a change in the profile of the type of workforce.

Another reason is a late government stance in promoting the qualification of labor to service the market, which has warmed up without labor to supply the demand.

## High Turnover of Labor

The high turnover is directly linked to labor shortages, since the law of "who offers more" has prevailed, and not only the salary itself, basic benefits, bonuses and incentives have prevailed, such as improvements in the basic basket, has been enough for workers to change businesses.

## Risk of High Accident Due to Hostile Environment

After few seconds thinking about the environment of the construction and assembly enterprises, we easily find that the high risk of accidents is a fact. Some tons being transported by winches, excavations, drilling machines and especially people with misperceptions about safety.

Despite security programs, which are increasingly publicized by companies and demanded by the government, we still have a large contingent of people who believe that by using security equipment they are doing a favor for the company, not caring for themselves.

Excessive overtime, delayed payments, poorly maintained comfort and hygienic housing, poorly designed beds, unskilled labor, also contribute to the concept of a hostile environment and, consequently, an increased risk of accidents.

## Significant Number of People and Companies Involved

Each company uses a driving strategy for construction and assembly of enterprises, but invariably, this is the stage with the largest number of contingents in the field. This in practice means strong coordination of field coordination, so that the objectives are fulfilled.

This factor becomes more relevant in projects which are developed simultaneously in several places in the world.

## Relevant Stakeholder Interference

The two preceding items culminate in this one, which means that there is a possibility that the enterprise may have its goals not achieved, for example, due to strikes and other events. Several stakeholders have power and authority to influence how the project will be developed and its outcomes.

## High Impact at the Site of Deployment

Whether it be constructing a building, a bridge, or a new industrial shed, none of them goes unnoticed in the mean they will be implanted.

This attention goes beyond public agencies, whether state or federal, political, legal and collection issues, because we have the attention of the local community, which aims minimum impact to their routine, but want improvements and development in the local economy.

## Use of Technology, Often Pioneering

A few minutes thinking, and although some techniques remain the same, as in the last century, others, however, do not even remember their past, due to the great competitiveness in the market, driven by innovation.

Every day there is a completely new product, or an old product in new format or material for better application.

Pioneering requires risk-taking and a workforce capable of carrying forward these characteristics, which are increasingly common in construction and assembly projects.

### Sometimes Embedded in a Political Setting

A building may have no political interest, but probably the release for its construction in a new area of the city does. The construction of large enterprises, such as stadiums, power plants, port expansion, undoubtedly have in their scenario a great political interest, for all the impact that will cause socially and economically in their place of implantation.

## CONCLUSION

Engineering and Construction projects require detailed planning and constant monitoring. Its peculiar features charge well-prepared staff, high sums of available financial resources and knowledge of the business and more comprehensive legislation (local and global). The realization and delivery of these projects is a challenge for experienced project managers.

## REFERENCES

Coutinho, I. (2009). *Study of the adherence of project management processes in consulting engineering companies in Belo Horizonte* (Master's Dissertation). FUMEC.

Disnmore, P., & Silveira Neto, F. (2004). *Project Management, How to Manage Your Project with Quality, within the Time Frame and Expected Costs*. Rio de Janeiro: Qualitymark.

Kerzner, H. (2003). *Project Management: a system approach to planning, scheduling, and controlling* (8th ed.). John Wiley & Sons.

PMI. (2017). PMBOK - Project Management Body Of Knowledge (6th ed.). Project Management Institute.

# Chapter 2
# Project Management and Maturity in Complex Engineering Projects


## ABSTRACT

*In this chapter, the authors present the main concept of project management and its influence on maturity within complex engineering projects. From the concepts of project management, the authors propose a model that will be discussed in the next chapters. Analyzing the main structure of the most applicable methodologies of the market today, they identify the minimum core that would be desired in companies with complex projects in order to ensure a better implementation of the portfolio. Project management is a pillar of support so that organizations can minimize their risks in investments, expansions, and operational improvements. As a result, companies are starting to rely on project management methodologies as a way to make their investments effective.*


## INTRODUCTION

PMBOK - Project Management Body of Knowledge - (2009) or Project Management Guide defines a project as a temporary effort undertaken to create a unique product, service or result. This way, project management appears as a pillar of support so that organizations can minimize their risks in investments, expansions and operational improvements. As a result, companies

DOI: 10.4018/978-1-5225-5864-4.ch002

are starting to rely on project management methodologies as a way to make their investments effective.

According to IPA - Independent Project Analysis - (2004), using FEL - Front-End Loading - its methodology for project management, is possible for the company to detail the scope of the project in order to achieve its objectives business.

The situation of project management becomes more complex when we think, for example, of large investments such as those in the mining industry. For example, in Brazil, according to the survey conducted by IBRAM - Brazilian Mining Institute (2011), investments in the mining sector showed that mining companies will have an investment capacity of approximately US $ 64.8 billion in 2011 to 2015 and that this could remain stable for the next few years. According to IBRAM (2011), investments between 2008 and 2012 were in the order of US $ 57 billion.

The economic crisis forced the investment forecast to fall to US $ 47 billion (period 2009-2013), according to the March 2009 survey. However, this sector is a great example of the importance of project management as a source of assurance and Capital economy.

From this perspective, DNPM - National Department of Mineral Production - indicates the rise of investments in mining:

*The Mineral Economy, in particular, energized by the recovery of mineral commodities, goes through a real euphoria, drawing a new mining-economic boom, with highly positive effects on the increase of net foreign direct investments, bringing greater reliability to the investors - attracted by the stability of the National economy, as evidenced by the successive falls in the country risk index (215 points, 01.03.2006). (DNPM, 2006, p.14).*

In this sense, for the implementation of complex projects, such as the ones mentioned above, it is necessary to have a specific project management methodology capable of ensuring the best allocation of resources in this type of investment. For this, the more mature the company is, the more competitive it becomes in allocating its resources.

According to Kerzner (2005), the maturity in project management consists in the development of processes that, by their very nature, become repetitive. With this, the probability of succeeding increases. However, success cannot be guaranteed.

As companies implement their methods and practices in project management, their maturity level grows as a direct effect to their adoption (PRADO, 2008). Thus, is needed a way to evaluate how mature companies are in relation to project management as a source for companies to identify their strengths and weaknesses, thus constituting the process of continuous improvement.

For complex engineering projects and investments, such as mining projects, there are still factors that are critical to the success of these projects such as: environmental management, stakeholder management, engineering and geology management, which need to be measured and analyzed, since they can affect the success of the project.

For instance, in some countries, as an example used previously, mining is one of the basic sectors of the economy of the country, contributing decisively to the well-being and improvement of the quality of life of present and future generations. It is fundamental for the development of an equitable society, provided it is operated with social responsibility, being always present the precepts of sustainable development (Wagner et al., 2002).

As the mining sector is characterized by investment, and these investments are materialized in several projects in the industry, it can be expected that the control and evaluation of the projects reflect how profitable and lucrative the sector can be.

In face of this investment scenario, complex engineering projects should focus efforts on the search for technologies, procedures, methods or practices that enable the management of their projects in order to assure goals and profitability. Searching for methods and practices of project management may reflect that organizations seek to learn, to know and to disseminate information, which can be understood as a model of maturity.

## PROJECT MANAGEMENT: STRATEGY AND PLANNING

For Coutinho (2009), once industries reached their historic mark of using their installation capacity, began to need new investments, with large projects emerging in several sectors such as food, agribusiness, fuels, mining, steel, etc. (Coutinho, 2009).

Success or failure in a duel, be it in business or sport, is intrinsically linked to the strategy developed by the party and the ability to plan and predict in advance, the actions of the rival. The word strategy has arisen in the past with the great battles and confrontations, according to Carvalho and Laurindo (2007). In industrial firms, things are no different, since in general they compete in their markets for greater and better results of operational and commercial growth through strategic planning and management.

Stalk and Houl (2002) argue that the 1980s were the decade of quality, and 10 years later, in 1990, the decade of responsiveness would be lived (the market produces a rapid response to customers). It is also possible to analyze this mechanism of rapid response directed at the need to expand and create new projects, according to Prado (2004), which compares companies to the human body, in the sense of reacting to external stimuli and, in addition, produce proactivity.

The way companies find to predict these changes and new business strategies is Strategic Planning. This planning, coupled with modern project management techniques, will stimulate and chart business growth around the world. Most of the time, the definition or option for the expansion of an organization, and the consequent origin of a new project comes from scenarios.

It is from this context that the Strategic Planning arises that can be understood as the alignment of an entire organization that will be focused on fulfilling a mission, aligned with the objectives of a vision. Ansoff (1977) defines Strategic Planning as the set of processes that will guide organizations in their decision making. According to the author, Strategic Planning leads to Strategic Management, which is the process of making the organization capable of integrating administrative and operational decisions with strategic decisions, while at the same time giving greater efficiency and effectiveness to the organization.

When the Strategic Planning is deployed in small components, in the end, it's obtained what is called Strategic Objectives. In order to fulfill each strategic objective, the company assigns a project to achieve this objective. Thus, through Strategic Planning, the portfolio of projects that the company must manage in order to achieve its objectives and to promote its growth emerges.

*Projects are often used as a means of achieving the strategic planning of an organization. Projects are usually authorized as a result of one or more of the following strategic considerations: market demand, opportunity or strategic business need, customer request, technological advancement, legal requirement (PMBOK, 2009, p.42).*

According to the PMBOK (2009), the strategic planning of an organization becomes the main factor of orientation for investments in the company's projects. In this way, it is concluded that, for the most part, projects are born from the need for investments that are guided by the strategic management coming from the strategic planning of the organization.

Competition is established in the industry when the specific sector presents more than one player in common. In this case, what strengthens competition is basically perceived when rivals try to improve their position or feel somehow pressured by the other rival.

This happens because most of the times, when taking an action, it can have repercussion in other companies, causing what we call interdependence. This phenomenon forces rivals to seek alternatives that enable competitors to overcome them, because if they do not, they may suffer impacts caused by these actions, which may become harmful and, in some cases, irreversible. In this way, competitors and participants in this industry are increasingly obliged to develop and sustain competitive actions and differentials so that they can always remain "alive" in their niche.

The term competitive advantage, strongly espoused by Michael Porter in 1986, is basically supported on the differentials that organizations must maintain in order to stay healthy relative to other competitors, promoting their profitability and advancement in the industry. In this same context, the author cites three fundamental strategies in this way so that companies can achieve this scenario: always lead by focusing on cost reduction, always promoting differentiation and always maintaining focus (Porter, 1986).

Porter (1986) also points out that for product development to become standard in the market, it is paramount that shows levels of differentiation capable of promoting competition with its main competitors.

Thus, the differentiation factor induces companies to try to achieve this competitive advantage through alternatives that are valued by consumers, although they present high costs to be produced. In this context, marketing is the main instrument for obtaining competitive advantage through differentiation.

If the generic strategy chosen by the company is the focus, it should focus on serving a specific niche market, and alienation from serving other segments of the industry as a whole. Once you choose the segment, it must combine focus strategy with cost leadership, or differentiation, to gain competitive advantage.

However, although Porter presents the three forms of strategy, the author also presents two points of concern: first, the company does not support the chosen strategy and secondly, the benefits achieved through the chosen alternative will be devastated by the growth of the sector (Porter, 1986, p.58).

According to D'Aveni (1995), when one approaches the static strategy, it is necessary to consider four competition spheres: "cost and quality", "timing and know-how", (D'Aveni, 1995). Still, D'Aveni (1995) considered that whatever competitive advantage may be developed it is considered fleeting.

## ENGINEERING PROJECT MANAGEMENT: IMPORTANT CONCEPTS

The way companies put their ideas in practice, potentiates the execution of the projects considered primordial for its success. According to the PMBOK (2009), a project is defined as follows:

*Project is a temporary enterprise, with defined start and end dates, with a purpose to create a unique product and service that is completed when its goals and objectives are reached and approved by stakeholders (PMBOK, 2009: 34).*

On the other hand, Kerzner (2003) defines the project as an enterprise with an identifiable objective, that consumes resources and operates under pressures of terms, costs and quality, and is generally considered as a unique activity of a company.

The projects are composed of a series of actions, called activities that form a chain with predecessor activities and successor activities to each one, originating an end, a network of activities, to which we call the project network. Typically, the activities of a project are grouped into phases that together give rise to what we call the life cycle. The life cycle varies from organization to organization, according to each specificity in the application of projects. According to Shtub, Bard and Globerson (1994), the phases of a project may be different, but generally are composed of:

- **Conceptual Project:** First step of the organization, where the need for the project is identified or a project is created;
- **Advanced Development:** Project planning and budgeting phase;
- **Detailed Project:** When the project is approved, the detailing is started;

- **Production:** Consists in the execution of what was planned;
- **Completion:** Last phase of the project.

According to the PMBOK guide (2009), a project comes to an end when at least one of the following is true:

- The objectives were achieved;
- It is clear that objectives will not or cannot be achieved;
- There is no longer a need for the project.

According to Heldman (2006), the concept of project management refers to the application of a set of knowledge, skills, tools and techniques directly to the activities of a project, with the sole objective of achieving the requirements for which it was designed. The application of this set of tools and techniques on a set of inputs from other processes produces what we call outputs processes of the project management, which allows the execution of project management throughout its cycle: initiation, planning, control and closure.

According to the PMBOK guide (2009), Project Management is defined as a coordinated set of activities whose main objective is to reach the expectations of the direct and indirect stakeholders, positively or negatively, during or after the conclusion of the project.

For Disnmore and Neto (2004), project management can be divided into three integrated dimensions. In its first dimension are the elements of the project and define what should be done. In the second dimension are the factors that specify the levels of project performance, such as cost, time and quality. And, in the third dimension, there are the tools for coordinating the work within the limits of the project, being: planning, control and evaluation.

Prado (2004) divides project management into basically three historical phases, as follows:

1. **Traditional:** Basically, only three variables were considered in its management process: deadlines, costs and quality;
2. **Modern:** It was based on customer satisfaction as the main success factor of a project and later on, the satisfaction of the project team members. In addition to these two factors, the following factors became important: deadlines, costs, quality, scope, human resources, communications, risks, supplies, contract and project integration;
3. **Corporate:** This is a recent way in the project management process. It is included in it a macro tracking of all the projects of the industry.

Implementation of the project management process within an organization involves the need to understand what is excellence in project management and what are the key factors that lead to its achievement. According to Kerzner (2005), in order to achieve excellence in project management, one must seek the use of a methodological form with great worldwide acceptance throughout the organization, as represented in Figure 1.

For Kerzner (2006), the most important is to make the processes integrated, in order to achieve excellence. It is intended, therefore, to approach the integration of project management with other administrative processes, such as total quality management (TQM), simultaneous engineering, reengineering, change management, risk management, among others.

*Figure 1. The 6 components of excellence in project management*
*Source: Kerzner, 2005*

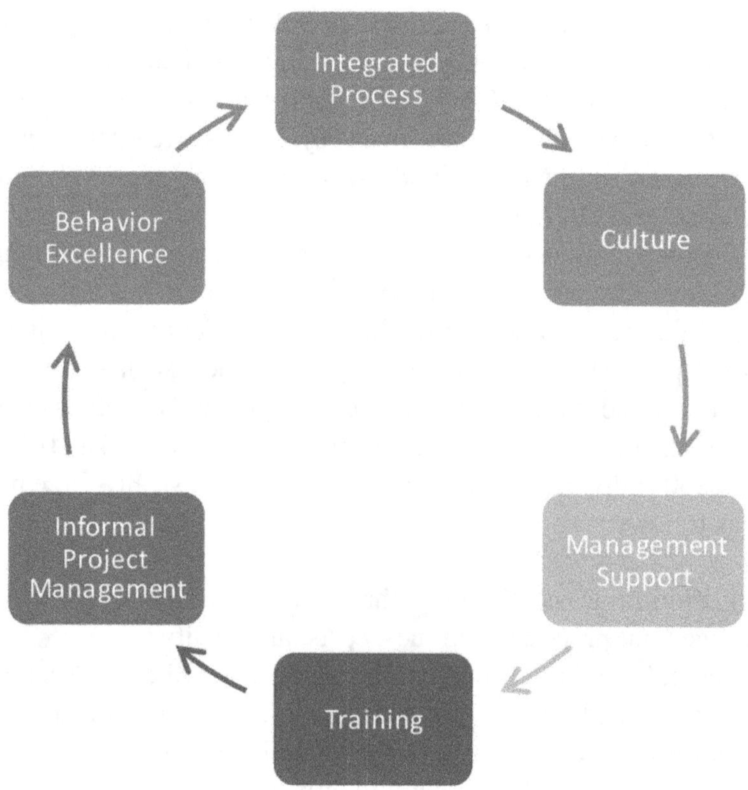

## ENGINEERING PROJECT MANAGEMENT: THE PROJECT MANAGEMENT OFFICE – PMO AND METHODOLOGIES

During the execution of the project, the project manager focuses and is very concerned about the execution of the project, so that the accomplishment is as close as possible to the planned one. To support the generation of reports and training, emerges what we call PMO - Project Management Office or Project Office.

*A project office is a body or organizational entity which is assigned various responsibilities related to the centralized and coordinated management of the projects under its control. The responsibilities of a PMO can range from providing project management support functions to being directly responsible for managing the project. (PMBOK, 2009, p.11).*

In this way, a company's PMO can be configured in the way that best fits the characteristic of this organization. As an example, a PMO with an advisory character is cited, that is, this PMO has the function of providing methods, practices, training, etc. for the project team while the project is being carried out. Thus, all project management is under the responsibility of its manager, who remains focused on its execution.

Within a general context, the Project Office is increasingly used in today's industries as a way to decentralize the project team's support activities, keeping them focused on direct management of that investment. According to the PMBOK (2009), while the project manager focuses on the project objectives, the PMO manages the scope change requests, and while the project manager controls the resources assigned to the project, the PMO optimizes the use of shared resources, among other activities.

The word "methodology" basically means a set of methods that form a guideline for the attainment of a goal. In project management, a methodology is composed of a series of steps to be followed, aiming to ensure the correct application of the processes. In short, it's what's called a "cake recipe" that an organization needs to manage its projects. Generally, the steps or processes of a methodology are composed of standardized tools and techniques to be used throughout the organization.

*Since people are the ones who do the work, it is essential that they do it the best they can and then the behavioral and contextual skills are equally important. Aspects of human relationships affect all involved in the projects. Innumerous negative daily conflicts, which end up harming the company in some way, could be avoided by knowledge of these aspects. Contextual competence is necessary because the project manager and others involved need to know the organization where they work and other organizations that may affect their project. They need to know financial and legal aspects. (Coutinho, 2009, p.56)*

Based on this context of methodology, we propose the evaluation of two methodologies in project management: PMBOK - Project Management Body of Knowledge of PMI - Project Management Institute and FEL - Front End Loading of IPA - Independent Project Analysis.

## *PMBOK:* Project Management Body of Knowledge

The acronym PMI comes from the expression Project Management Institute, or Institute of Project Management, non-profit, exclusively focused on project management. Based in Philadelphia, United States, the institution was founded in 1969 and is now considered a global community with more than three hundred thousand participants and present in more than 160 countries.

Its affiliates come from different areas such as: naval, mining, information technology, etc. Its main goal is to promote project management in a standardized manner throughout the world. As a leading product, PMI manages the issuance of PMBOK - Project Management Body of Knowledge.

Besides the PMBOK edition, PMI manages several other standards, as well as several certifications in knowledge in the field of project management, among which PMP - Project Management Professional stands out.

The PMBOK (Best Practice in Project Management) consists of a repository of project management tools and techniques that are applied as the project progresses in its phases.

*The Project Management Knowledge Guide (PMBOK) is a recognized standard for project management. A standard is a document that describes established patterns, methods, processes, and practices. As in other professions such as advocacy, medicine and accounting, the knowledge contained in this standard evolved from the recognized best practices of project management professionals who contributed to its development (PMBOK, 2009, p.3)*

As the PMBOK (2009) states, it is a guide and not a methodology. It serves as a basic reference for consultation to the best tool to be applied in a given phase of the project. Thus, each company discusses and implements its structure in the way that best suits the type of project, industry, scenario and, in general, the peculiar characteristics that each project has.

*The PMBOK Guide identifies the broadly recognized set of management knowledge as good practice. "Widely Recognized" means that the knowledge and practices described are applicable to most projects most of the time and that there is a consensus as to its value and usefulness. "Good Practice" means there is a general consensus that the correct application of these skills, tools and techniques can increase the chances of success in a wide range of projects. Good practice does not mean that the knowledge described should always be applied uniformly in all cases; the organization and / or project management team is responsible for determining what is appropriate for a particular project. (PMBOK, 2009, p.4)*

The PMBOK does not propose a life cycle for projects. On the contrary, the PMBOK (2009) defines the life cycle of a project placing the necessary phases to execute this project. In Figure 2, is exemplified a lifecycle model and the behavior of resources throughout project execution.

*The life cycle can be defined or shaped according to the unique aspects of the organization, industry or technology employed. While all projects have a defined beginning and end, deliveries and specific activities conducted in the meantime may vary greatly depending on the project. The life cycle provides a basic framework for project management, regardless of the specific work involved. (PMBOK, 2009, p.15)*

For the execution of the phases of the life cycle of a project, resources (equipment, materials and labor) are necessary to fulfill each stage of the project. Labor resources, for the most part, are organized into organizational structures.

According to the PMBOK (2009), these structures can vary from functional to project oriented depending on the decision power of the project manager.

Basically, the PMBOK is structured as follows:

- Chapters 1 and 2: introduces the concept of project management;

*Figure 2. Level of resources during the project life cycle*
*Source: Project Management: A Managerial Approach, 2010*

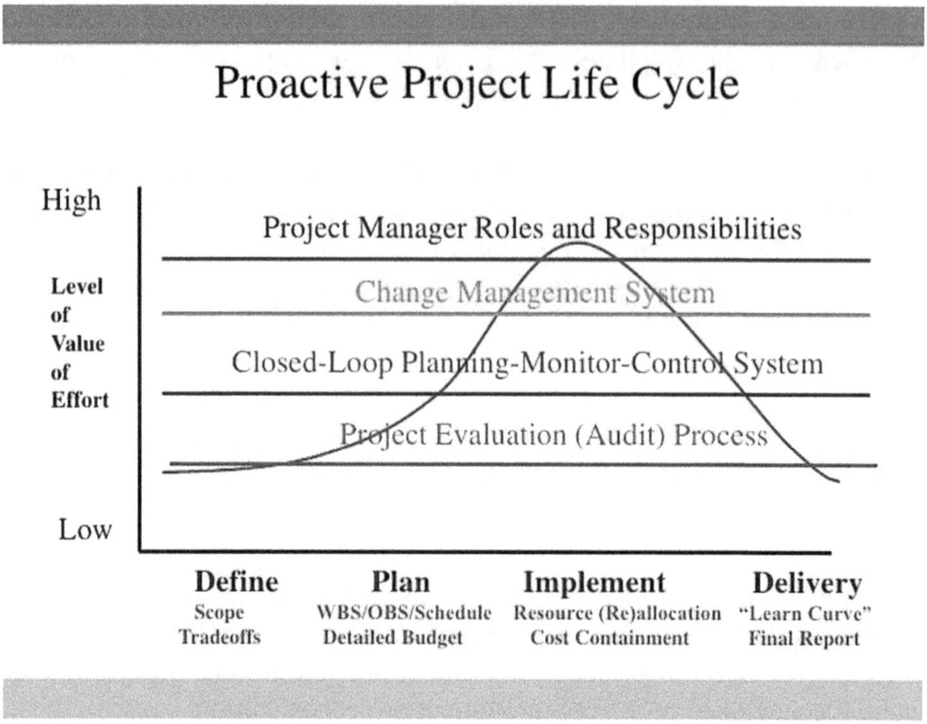


- Chapter 3: Describes the project management pattern. It summarizes the processes, inputs and outputs that are good practices in several projects;
- Chapters 4 through 12: Describes the guide in project management.

The PMBOK organizes its knowledge and divides it into 42 separate processes into five groups:

- **Initiation:** consists of the processes that formalize the beginning of the project or a project phase, such as documenting the interests of all those involved in the project. The main documents produced in this phase are the Opening Term or Project Charter, a document that officializes the beginning of the project and names the project manager, in addition to the Stakeholder Registry, which documents all the requirements of those involved or impacted by the project;

- **Planning:** as the name says, it is in this process that the scope of the project, budget, deadline, communication, human resources, risks and supplies are planned, which, in the end, together constitute the main document of the project: Project Management Plan. In this area, the Project Management Plan is the main document of this process, which will guide its execution;
- **Execution:** in this process, the Project Management Plan produced in the Planning Process is executed. Its main function is to produce planned deliverables within the cost, scope, and time frame contained in the Project Management Plan. It is also observed that in addition to the execution, in this process, these are requests for change, quality control of deliveries, etc.;
- **Monitoring and Control:** In the Monitoring and Control process there is the monitoring and management of everything that was planned and if these steps are being executed according to the Project Management Plan. This process encompasses all other processes as monitoring and control takes place throughout the project life cycle;
- **Closure:** it is in this process that the project phases are formally closed, deliveries are evaluated, contracts with suppliers are closed and the project is finalized. Besides these activities, the lessons learned during the project are documented so that strengths are maximized in future projects and weaknesses are avoided.

In addition to the five process groups, in which the 42 processes are organized, the PMBOK (2009) presents these processes organized into nine areas of knowledge as shown in the sequence. This separation by the five processes or the nine areas of knowledge can be clearly understood as:

- **Integration:** Corresponds to the processes that promote the integration of the project as a whole;
- **Scope:** Groups the processes related to the development and management of the project scope;
- **Time:** Understands the processes related to the production of the project schedule;
- **Cost:** Corresponds to the processes responsible for preparing the project budget;
- **Human Resources:** Set of processes related to human resources of the project, organizational structure, team management, etc.;

- **Communication:** Groups the processes related to the production, management and submission of project information to the team;
- **Risks:** It includes the processes of planning and evaluating the risks of the project;
- **Quality:** Corresponds to the processes necessary to guarantee the quality of both the project and the product of the project;
- **Procurement:** Set of processes related to the planning and contracting of equipment and services for the project.

Each process of the 42 processes are composed of entries, which are documents captured during the project. These documents are worked by tools and techniques of each process and produce outputs that are products and documents that will serve as inputs to other processes. In this way, the processes interact throughout the project.

## FEL: Front End Loading

Independent Project Analysis (IPA) is an independent institute founded in 1987 for the sole purpose of providing project management research for the chemical, oil and mining industry. According to the Research and Metrics Measuring Capital Project (2004), the institute's researchers are mostly highly educated engineers, economists, scientists, statisticians, and operational researchers. IPA has offices in the United States, the Netherlands, China, Australia and recently opened an office in Curitiba, Brazil.

Its main mission is to improve the competitiveness of its clients by improving and optimizing the management of its investment projects by comparing and evaluating the maturity of projects compared to similar projects contained in its database.

Unlike the PMBOK, Front End Loading (FEL) is a methodology widely disseminated by the IPA, as its own and characteristic life cycle, containing all necessary steps, from the beginning of the project to its completion. Based on the Research and Metrics Measuring Capital Project (2004), the phases of its life cycle evolve as project requirements are known to ensure cost and schedule planning. In Figure 3, it's observed the evolution of its life cycle, as the project begins to be detailed.

The FEL methodology is constituted by a set of processes widely used in mega projects, technically called capital projects. In most cases, these projects require large investments and the FEL methodology is used to minimize the risks of non-viable and unattractive investments for the organization.

*Figure 3. FEL life cycle project*
*Source: Project LifeCycle Process Engineering • Consulting • Technologies Project And Construction Management, 2008.*

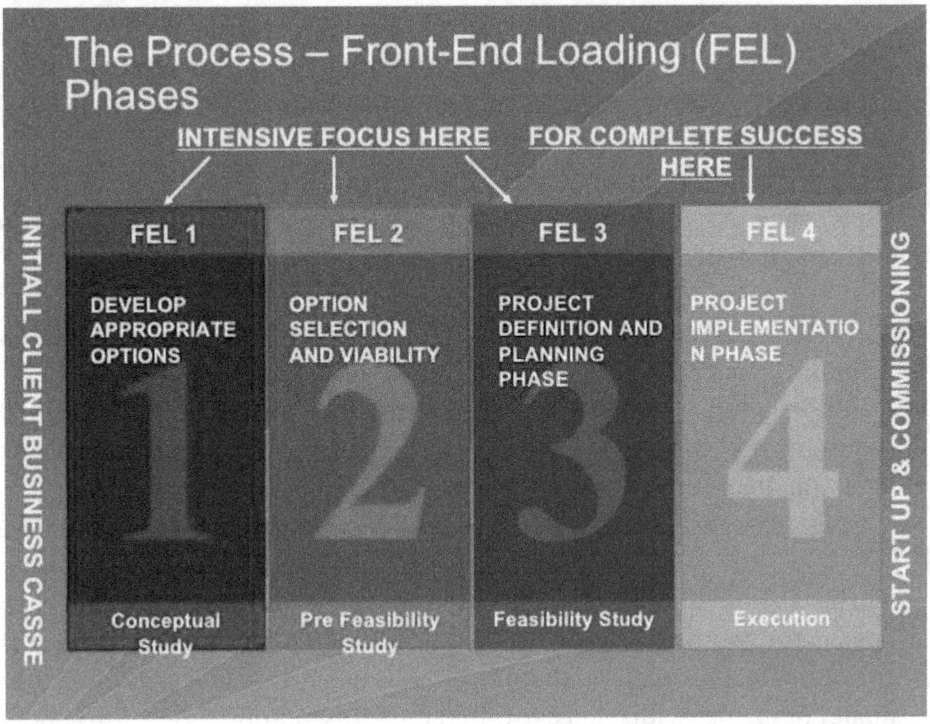


Typically, FEL is used in the industrial sector, such as mining, energy and petrochemicals, in which projects are highly complex and costly.

For the transition between the FEL phases, for example, from FEL1 to FEL2, there is what is called Evaluation Gates, which require approval from the company's executive body to decide whether the project continues or is discontinued, or even to ask for more information for decision making.

According to Figure 3, the FEL – *Front End Loading* cycle includes the processes:

- Conceptual Study, which covers processes for analyzing alternatives to the project, financial calculation of investments, such as financial return, etc. This process comprises the FEL1 phase;
- Pre Feasibility Study, which consists of a vision of the concept of the project before proceeding to a later stage in which the financial resources of the project begin to be spent. In this case, this process represents the FEL2 phase;

- Feasibility Study or Project Definition and Basic Planning, which represents the FEL3 phase, from which the project, already in its final version for approval, is presented to the executive body of the organization before the execution of the project begins .
- Execution or Implementation of the Project, which represents the FEL4 phase, from which the project, already approved, is implemented according to the previous approved premises.

Thus, once the phase FEL3 is approved, the execution of the project in the FEL4 is initiated through the processes of:

- Detailed Engineering, which represents the detailed engineering design for the construction;
- Construction, which includes the construction and execution of detailed design engineering;
- Startup and Turnover, when the product of the project is started;
- Operations, which includes the start of operation of the project product.

Thus, as a goal of this book, the development of the book will be through critical maturity processes, such as stakeholder planning, analysis of existing maturity models, as well as the proposal to use a new maturity model to evaluate companies with intensive Capital, such as mining, civil construction, energy and others.

In this way, the purpose of this book is to introduce the reader to the basic concepts of project management, main books and methods used today in the process of managing complex projects and to propose through this, a new way of approaching complex projects and analyzing specific pillars and relevant only to that project niche.

## CONCLUSION

After the presentation of this chapter, now the reader understood the main and best used practices in projects management. It's important to mention that, in the industries today, given the singularity of each one of them, usually they customize their own methodology. In others words, to better attend the need of the industry and more than this, ensure that it's covering the values of that industry, they choose between of the best project management practices, the ones that better fit in the need of that industry.

This is also considered an important step of the industry in the way to pursue the maturity and the desired level of project management. It includes all the process to map: the lifecycle of the projects executed by the industry and what tools from the pool will fit in the methodology proposed.

The reader also needs to understand that, the objective of this book is to introduce a new concept of maturity measurement adapted in order to include and to map the different points and crucial ones that this kind of project brings to the organization. The factors are, for example, when talking about mining, we should understand: the geology point, the environmental licensing, etc.

## REFERENCES

Ansoff, I. (1977). *Business Strategy.* São Paulo: McGraw-Hill.

Carvalho, M. M., & Laurindo, F. J. B. (2007). *Competitive Strategy: from concepts to implementation. 2.* São Paulo: Atlas.

Coutinho, I. (2009). *Study of the adherence of project management processes in consulting engineering companies in Belo Horizonte* (Master's Dissertation). FUMEC.

Disnmore, P., & Neto, S. F. (2004). Project Management, How to Manage Your Project with Quality, within the Time Frame and Expected Costs. Rio de Janeiro: Qualitymark.

DNPM - Departamento Nacional De Produção Mineral. (2006). *Mineral Business: Investor's Guide in Brazil.* Available: http://www.ibram.org.br/150/15001005.asp?ttCD_CHAVE=21653

Heldman, K. (2006). *PMP: Project Management Professional – Study Guide. 3.* Wiley Publishing.

IBRAM - Instituto Brasileiro De Mineração. (2009). *The Mining Industry in Minas Gerais.* Available: http://www.ibram.org.br/sites/1300/1382/00000177.pdf

IBRAM - Instituto Brasileiro De Mineração. (2011). *Information and Analysis of the Brazilian Mineral Economy.* Disponível em: http://www.ibram.org.br/

Independent Project Analysis (IPA). (2004). *Research And Metrics Measuring Capital Project.* Author.

Kerzner, H. (2003). *Project Management: a system approach to planning, scheduling, and controlling. 8*. John Wiley & Sons.

Kerzner, H. (2005). *Using the Project Management Maturity Model – Strategy Planning for Project Management. 2*. John Wiley & Sons.

Kerzner, H. (2006). *Project Management: Best Practices. 2*. Porto Alegre: Bookman.

PMBOK - Project Management Body Of Knowledge. (2009). *PMI - Project Management Institute. 4*. PMI.

Porter, M. E. (1986). *Competitive strategy: techniques for analyzing industries and competition*. Rio de Janeiro: Campus.

Prado, D. (2004). *Project Planning and Control*. Belo Horizonte: INDG TECs.

Prado, D. (2008). *Maturity in Project Management*. Belo Horizonte: INDG TECs.

Project LifeCycle Process. (2008). *Engineering, Consulting, Technologies Project end Construction Management*. Available: https://pt.slideshare.net/bobotang/pl-pr3-presentation

Project Management. A Managerial Approach. (2010). *PM 01 - Introduction to Project Management*. Available: https://www.slideshare.net/ianachmadjanuar/pm-01-introduction-to-project-management

Shtub, A., Bard, J. F., & Globerson, S. (1994). *Project management engineering, technology and implementation*. Prentice-Hall.

Wagner, A. (2002). *The presidential election and mining*. Gazeta Mercantil.

# Chapter 3
# Culture, Knowledge Management, and Maturity in Complex Engineering Projects


## ABSTRACT

*This chapter opens a discussion of how mature organizations in project management deal with culture and knowledge management in complex projects. This discussion also includes organizational structure and communication aspects. This chapter seeks to establish relationships between these four aspects and the three main characteristics of complex projects: ambiguities, multiple stakeholders, and multinationality. At the end of chapter, a table summarizes these relationships. The purpose of this chapter is to begin a discussion of the culture and knowledge management aspects of mature organizations dealing with complex projects. Along with culture and knowledge management, the authors also cover aspects of organizational structure (for project implementation) and communication, since they have a strong relationship with each other. In fact, the boundaries between these four aspects (culture, knowledge management, structure, and communication) are not clear and there are overlaps between them.*


DOI: 10.4018/978-1-5225-5864-4.ch003

## INTRODUCTION

The purpose of this chapter is to begin a discussion of the culture and knowledge management aspects of mature organizations dealing with complex projects. Along with culture and knowledge management we will also cover aspects of organizational structure (for project implementation) and communication, since they have a strong relationship with each other. In fact, we can say that the boundaries between these four aspects (culture, knowledge management, structure and communication) are not clear and that there are overlaps between them.

Let's highlight some concepts related to complex projects that will be used in this chapter.

Increasing the pace of change, growing economic complexity, and global competition require executives to re-examine their strategies to meet stakeholder expectations and meet market needs. It is in this context that maturity in project management fits.

The Project Management Institute (PMI, 2013a) has published a specific edition of complex projects called "Navigating Complexity" in one of its "Pulse of profession in-depth reports". In it, which also addresses this context, PMI indicates the reasons that make it highly relevant to discuss complexity in projects. He says that complexity tends to grow larger and that it is usually associated with significant budgets. Thus, knowing how to deal with complexity becomes a competitive advantage for companies.

In the same report, PMI (2013a) lists the eleven main characteristics that define a project as complex, as follows:

1. Multiple stakeholders;
2. Ambiguity of project features, resources, phases, etc;
3. Significant political/authority influence;
4. Unknown project features, resources, phase, etc;
5. Dynamic (changing) project governance;
6. Significant external influences;
7. Use of a technology that is new to the organization;
8. Use of a technology that has not yet been fully developed;
9. Significant internal interpersonal or social influences;
10. Highly regulated environment;
11. Project duration exceeds the cycle of relevant technologies.

Many of these characteristics are directly related to the first two: ambiguity and multiple stakeholders. Thus, for our discussion, complex projects are those that have ambiguities and multiple stakeholders.

On ambiguities, it is convenient to highlight that they are conceptually related to enterprise environmental factors, which are input to several of the processes proposed by the PMBOK (PMI, 2013c), which are also related to the concept of risks. Concerning stakeholders, it should be mentioned that different stakeholders may have different expectations, which can compete with each other and create conflicts within the project. Our discussion will also consider the combination of these two characteristics, that is, projects that have stakeholders with ambiguous positions.

A third characteristic that we will consider in our discussions of complex projects is multinationality. Multinationality is understood as the characteristic of projects whose activities are dispersed in diverse places of the globe, something usual in our globalized world. Multinationality can make the issue of (multiple) stakeholders more complex by adding the ingredient of potentially meaningful cultural differences between them. These cultural differences can bring new ambiguities to the project.

An example of complex projects where stakeholders have ambiguous positions are those that have a significant impact on the communities where they are located, especially if they are small communities. Many mining projects or hydroelectric plant projects fall into this group. These are projects that, in addition to internal stakeholders, have a number of external stakeholders, such as:

- Public administration;
- Regulatory agencies;
- Non-governmental organizations (NGOs);
- Local merchants;
- Local workers;
- Residents of the community, among others.

Some of the ambiguities associated with these stakeholders are:

- Will public administration and regulatory agencies provide the necessary (environmental and other) licenses for the project to be deployed and operated?

- What is the position of the residents in relation to the project? Local merchants and workers can see some benefit in the implementation of the project, but what about the others?
- Can the conditions imposed by the public administration and regulatory agencies, based on the demands of the community, for the issuance of licenses be able to derail the project or to reduce its profitability significantly?
- Can events occur during the implementation of the project that stresses the relationship between the proprietary organization and the community?

Here are two quotes that show how important the above questions are. First, the consulting firm EY's report (2016) indicates that the fourth highest business risk in the mining and metals business in the biennium 2016-17 is related to the social license to operate (SLTO), that is, the agreement of the external stakeholders regarding the existence of the project. In this report, EY also states that "mining and metals companies need to embed sustainability into long-term planning and operate in tandem with local communities to retain social license to operate (SLTO)."

Secondly, Deloitte (2015) points to changes in how to dialogue with stakeholders, especially the external ones, as the sixth topic in a list of the 10 most important issues for mining today.

We will also make some references on project management maturity, which will be used in this chapter. Our reference models for discussion on maturity are four, as follows:

- Capability Maturity Model (CMM) developed by Software Engineering Institute (SEI), located at Carnegie-Mellon University in Pittsburgh.
- Organizational Project Management Maturity Model (OPM3) developed by PMI;
- Project Management Maturity Model (PMMM) developed by Harold Kerzner;
- Maturity Model in Project Management (MMGP) developed by Darci Prado.

Darci Prado is a Brazilian consultant which holds a PhD in engineering from UNICAMP, Brazil. The MMGP was developed from a research whose data collection was carried out from October to December 2005 through the website www.maturityresearch.com. This research was answered by more than 250 professionals of all types of Brazilian organizations. This original work has been constantly updated and MMGP is widely used in Brazil.

In general, these models establish levels of maturity. Usually they are five levels, denominated by numbers or names, according to the model. Thus, the term "maturity level" will be usual in our discussions.

We say that a company has mature project management when it has developed and uses systems and processes of a repetitive nature that guarantee a high probability of success. Thus, the word "mature" here means an environment where predictability is high and risk (the unknown) is low.

Here are two quotations that demonstrate the relationship between repetitive nature and probability of success.

Persse and Persse (2001) state that:

*Full maturity implies that the organization has in place practices, policies, and disciplines that allow the group to produce quality software in a predictable, reliable, and repeatable manner. The SEI estimates that only about 1 to 2 percent of organizations worldwide operate at Level 5.*

Level 5 is the highest level of maturity in the CMM model.

Secondly, PMI (2013b) states that:

*OPM is a strategy execution framework utilizing portfolio, program, and project management as well as organization-enabling practices to consistently and predictably deliver organization strategy producing better performance, better results, and a sustainable advantage.*

Maturity does not necessarily mean having success in the implementation of a project, understanding success as the attainment of scope, quality, time and cost demands, and also obtaining customer satisfaction. Despite the differences between maturity and success, the higher the maturity of an organization's project management, the more successful it will be.

The opposite is true as well. A low level of maturity has a direct relationship with failure to develop projects, as demonstrated by Prado (2015):

*A maturity model would be [...] a mechanism capable of numerically quantifying the ability of an organization to successfully manage projects ... [According to the MMGP model] the average maturity of Brazilian organizations in 2014 was 2.64 [On a scale of 1 to 5 and consequently] the total success rate for Brazilian organizations in 2014 was 56%.*

According to Kerzner (2006) in mature businesses:

- Project management has come to be seen as an instrument for success, no longer as an interference with day-to-day goals.
- The organizational changes in the structure promoted by project management led to the maximization of the use of human resources.
- Synergy was present in mature companies in project management.
- The entire company supported the project manager and their requirements.

His statements highlight the relationships between maturity in project management and the aspects of culture, knowledge management, structure and communication, which we will discuss in the sequel.

## COMPANY CULTURE

The culture of a company, and each company has its own culture, manifests itself in itseveralaspects: its beliefs (vision, mission and values), its structure, its style of leadership, its processes and practices, its relations with stakeholders and consequently in how project management is done and in the results obtained.

We have included "processes and practices" in the description above to emphasize that a company's maturity is obtained through systems and processes of a repetitive nature and is heavily influenced by culture.

Kerzner (2006) proposes four basic values in project management:

- Cooperation;
- Team work;
- Confidence;
- Efficient communications.

The key word here is cooperation, which nourishes and is nurtured by teamwork, trust, and efficient communication. A culture of cooperation gives the organization the ability to react more quickly to crises.

A manifestation of the company culture, which also makes present in the management of knowledge and communication, is the use of terms to deal with management issues. In other words, each company has its own vocabulary. Thus, it is convenient to standardize some terms between project management and other management activities of the company.

## Culture, Ambiguities and Stakeholders

Kerzner (2006) also presents a classification of business cultures, summarized in Table 1, which is adequate for our discussion.

Given the fact that complex projects are characterized by multiple stakeholders and that they, especially internal stakeholders, have to deal with the ambiguities of this type of project, we can conclude on the importance of a cooperative culture for companies that aim to reach management maturity of projects. In other words, co-operation between several internal stakeholders is critical to succeed in dealing with the ambiguities of a complex project.

It is convenient to clarify that internal stakeholders include, but are not limited to, the team in charge of the project implementation and more specifically of its management.

The greater the culture (and practices) of cooperation within a company the greater is the possibility of achieving a level of maturity in project management. Thus, culture is one of the most important characteristics, perhaps the most important, in companies with excellence and maturity in project management. Culture promotes behavior focused on success.

The development of a culture (of cooperation) within an organization is a potentially time-consuming process and the maintenance of the results demands constant attention because the gains can be lost quickly.

The first step in the development of culture is to identify the main actors that can influence this cultural change. In this context, senior management support is key. Thus, in an organization that seeks to establish and maintain a culture suited to mature project management, there is no room for the personal whims of someone from top management who refuses to support project management.

*Table 1. Types of business cultures*

| Culture | Main Characteristics |
|---|---|
| Cooperative | Efficient communication, trust and cooperation. |
| Non-cooperative | Incapacity of senior managers to work together. |
| Competitive | Teams of projects compete with each other for resources. |
| Isolated | Different units of the company develop their own culture. |
| Fragmented | Within the same team are different cultures. |

Source: Author based on Kerzner (2006)

## Culture and Multinational Projects

Multinational projects will typically have to deal with issues related to cultural differences, whether within the team working on the project implementation or across stakeholders. That is, in a globalized world, understanding the impact of cultural differences in the development of a project that encompasses diverse organizations or stakeholders scattered across different parts of the globe is essential to its success.

Thus, cultural differences make matters related to ambiguity and multiple stakeholders even more critical.

In this context, Kerzner (2006) states that:

*Perhaps the biggest challenge facing senior management in this decade is the creation of a multinational project management culture." Cultural change requires transformations, and in some countries such transformations are difficult to implement.*

Thus, it is necessary to develop a multinational culture and multicultural skills.

To conclude, we can say that the relationship between project management and the culture of the organization is so much that, more than a set of norms or procedures, project management is a manifestation of the company culture. In other words, project management is based on the behavior of the organization and not on procedures.

## STRUCTURE

Prado (2015) makes some relevant observations about the importance of the structure for project management:

*Because project execution often involves different sectors of an organization, there is a need to choose the appropriate organizational structure (or governance structure) to maximize results and minimize conflicts [...] The organizational structure should regulate the relationship of authority and power between project managers and the various areas of the organization involved with the projects.*

Organizations may be of a non-project oriented or, on the other side of the scale, may be project oriented.

The use of the word "scale" was intentional, since between these extremes there are a multitude of companies that mix, to varying degrees, a non-project-oriented characteristics with a project-oriented one, that is, there are hybrid companies.

Regardless of the type of company (not project oriented, hybrid or project oriented), the fact is that all of them seek to continuously increase their efficiency and profitability through the best use of their (always scarce) resources. Thus, the more appropriate the structure for a company, the better it will use its resources.

## Types of Structure

The PMBOK (PMI, 2013c) proposes functional structures and, on the other side of the scale, projectized structure. Functional structures are hierarchical structures, where subordination relationships are clear. In this type of structure, the team that works on the development of a project is distributed across several functional areas and coordination is also dispersed. See Figure 1.

The projectized structures group the teams by specialties, which can be subdivided by functionalities. In this type of structure, the team who works on a project works under a single coordination. See Figure 2.

Between the extremes of the functional and projectized structures we have a variety of intermediate structures called the matrix ones. If we consider that a great organization consists of several levels, the combinations within it of part functional structures and part projectized can occur in several ways.

The PMBOK (PMI, 2013c) divides the matrix structures into three groups: weak, balanced and strong matrix structure. The weak matrix structure is closer to the functional structure, in the "scale of structures", while the strong matrix is closer to the projectized one.

Naturally we associate non-project oriented companies with functional structures and companies of a project oriented nature with projectized structures. From this association we could assume that a mature company in project management has to be a project oriented company with a projectized structure. Even though this is true for many mature project management companies, it does not necessarily apply to all companies.

Kerzner (2006) makes an observation that challenges this common sense:

*Figure 1. Functional structure*
*Source: Author based on PMI (2013c)*

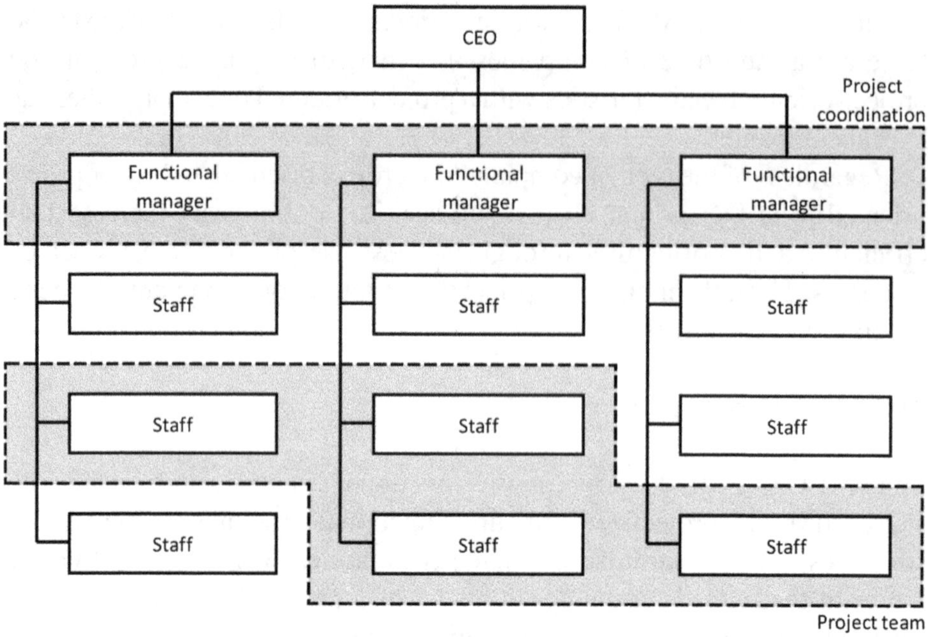


*Figure 2. Projectized structure*
*Source: Author based on PMI (2013c)*

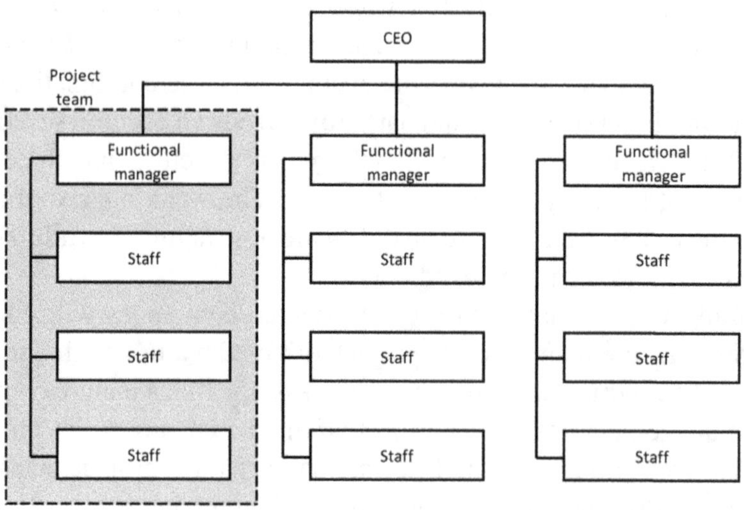

*Successful project management can happen within any structure, as bad it may seem on the project, as long as the company's internal culture is able to encourage teamwork, cooperation, trust and effective communication internally.*

That is, the culture of a company is a more relevant (or dominant) feature than definitions regarding its structure. Another way of thinking about Kerzner's words is to consider that the structure that will contribute to a company making the best use of its (scarce) resources is the one that is better aligned with the culture of the company. Obviously, in the context of our discussion, the culture in question must be appropriate to a mature project management, as discussed in the previous item.

## Projects, Programs, Portfolios and PMO

One of the aspects of mature project management that relates to the structure of companies is how these companies structure (gather and organize) their various projects.

It's up to us here to talk about program and portfolio project, so let's talk about Project Management Office (PMO).

If a project is a temporary effort made to create a single product or service, a program is a group of projects, subprograms, and related activities that are managed in a coordinated way to obtain benefits that could not be achieved if individually managed. For a program, centralized management and control allow both the operationalization of each one as well as the maintenance of the joint vision of its objectives.

A portfolio is a set of projects, programs, sub-portfolio and operations, which are managed as a group to achieve strategic objectives of the organization. This alignment between the portfolio and the organization's strategic objectives is its main feature.

The recognition that companies assemble and organize their projects in this way (programs and portfolios) is a fundamental part of maturity models. There are several portfolios and programs citations in maturity models. For example, PMI (2013b) states that: "OPM3 transforms the portfolio, program, and project domain processes into high-quality delivery processes that are well understood, stable, repeatable, and predictable."

As for the PMO, it is a management structure that standardizes the processes related to the project and facilitates the sharing of resources, methodologies, tools and techniques. It is also the link between a company's portfolios, programs and projects and corporate measurement systems, such as the balanced scorecard.

The way a PMO acts, and consequently the level of control it has over the development of an organization's projects, varies from one organization to another. We can relate three types of proceeding, which are usually cumulative, according to Table 2.

In line with the above types of performance, the PMBOK (PMI, 2013c) lists the following PMO functions to support project managers:

- Manage shared resources in all projects managed by the PMO.
- Identify and develop project management methodology, best practices and standards.
- Train, guide and supervise.
- Monitor compliance with project management standards, policies, procedures and models through project audits.
- Develop and manage shared project policies, procedures, models and other documentation (organizational process resources).
- Coordinate communication between projects.

If we consider that a company has maturity in project management when it has developed and uses systems and processes of a repetitive nature and that guarantee a high probability of success, we can conclude that PMO had and has a role of extreme relevance in this journey.

The PMO acts to ensure that practices are maturing, so that good practices emerge and so that the company can have good project governance. The

*Table 2. Types of performance of a PMO*

| Type | Degree of Control | Description |
|---|---|---|
| Supportive | Low | Advisory paper on projects. Provides models, best practices, training, access to information and lessons learned from other projects. |
| Controlling | Moderate | Provides support and requires compliance. It covers the adoption of project management structures or methodologies, using specific models, forms and tools, or compliance with governance. |
| Directive | High | Take control of the projects by directly managing the projects. |

Source: Author based on PMI (2013)

establishment of continuous improvement processes will contribute to the improvement of project management practices.

## Structure and Stakeholders

In a globalized world and with an increasing frequency of complex projects, it is unrealistic to think about developing projects with only teams internal to the company. So, let's talk here about a specific

The outsourcing of activities in the project implementation, such as detailed engineering and construction, began a process of reducing the size of the own teams of the proprietary organizations. Use of third-party is a reality in project management, that is, proprietary organizations are increasingly dependent on third parties as projects become more complex.

In this context, the teams include both staff from the owner organization and contracted third-parties. This reality means that:

- More than the name of the proprietary organization, the team must bear the name of the project;
- The mix of personnel from the owner organization and from (more than one) third party requires constant efforts and monitoring for team building;
- Demands specific actions in knowledge management.

The project team, its own and third parties must cover all the skills and functions that influence the outcome of the project, the role and responsibility of each member of this team must be defined. It is understood that "defined" covers the definition itself, but also that it must be communicated, understood and agreed upon.

Team members can be deployed full-time or part-time and can be added or removed from the team depending on project development. Regardless of this, the risk of unavailability of any member of the team, in fact, of this member's abilities, should be contemplated in the project planning.

Areas with potential problems (ambiguities and risks) should be defined, with specific plans developed and responsibilities assigned. Here we have an area of contact between the processes of human resources management and risk management in the development of the project.

## Structure, Ambiguities and Multinational Projects

For complex and multinational projects, two aspects should be considered.

First, the use of virtual teams, that is, the use of team whose members do not work or work for a limited period of time in the same place. There are many possibilities associated with this, possibilities directly related to the need to use the (scarce) human resources in an efficient and profitable way. Here are some examples, including examples that do not apply exclusively to multinationality:

- The team has specialists working in locations other than the project site.
- The team has professionals who work usually in locations with significantly different time zones.
- Use professionals who work in the home office mode.
- Use professionals who have some type of restriction or mobility impairment.
- Reduce travel expenses for team members.

Secondly, the laws regulating the work of third parties vary from country to country. Thus, a practice that is usual and licit in one location can generate labor liabilities in another. This type of question should be duly considered when defining the team that will work on a project, as well as other aspects of human resource management of this enterprise.

This is a good example of ambiguities involving stakeholders that characterize complex projects. Labor liabilities can generate additional costs or oppositions to the project by some stakeholders, regulatory agencies and the local community, for example. Thus, the company must be aware of these risks in order to be able to manage them in the best possible way.

## Leadership

We cannot talk about structure and staff without making some comments about project leadership.

Leaders, whether project managers, PMO managers or others, are strategic elements for the development and maintenance of the organization's culture and for the structure (of which they are part of) to function efficiently and profitably in project development. Thus, alternation in leadership roles throughout the project life cycle, when necessary or inevitable, demands care.

The leader (and management team) of a project must ensure that all team members follow professional and ethical behavior aligned with the company culture. They should also pay attention to shared responsibilities and rewards. This is especially true in companies where the structure is part functional and part projectized. All those who contributed to the success of the project, whether "functional" or "projectized", should be recognized for their contributions.

The leader of a project must have leadership, technical, and strategic skills (related to the organization's business). This will be explored in the next topic, Knowledge Management.

## KNOWLEDGE MANAGEMENT

Prado (2015) lists the seven dimensions of its maturity model (MMGP):

1. Competence in project management, programs and portfolios;
2. Behavioral competence;
3. Technical competence;
4. Use of methodologies;
5. Informatization;
6. Organizational structure; and
7. Strategic alignment.

The relationship of the first three dimensions, the competences, with knowledge management is direct, since the competences can be developed with training programs that are part of the knowledge management system of the company.

The PMI (2013b) states that OPM integrates the following factors:

- Knowledge (of the portfolio, program, and project processes);
- Organizational strategy (mission, vision, objectives, and goals);
- People (having competent resources);
- Processes (the application of the stages of process improvement).

Analyzing these factors, the importance of knowledge management emerges as one of the integration tools.

The citations above demonstrate the relevance that knowledge management has for companies that intend to have a high maturity level in project management.

We will discuss the management of knowledge on two aspects: first, the qualification of the leaderships and the teams of the projects and secondly, the maintenance of the knowledge that are relevant within the organization. In addition, we will explore its relationship with ambiguities, multiple stakeholders and multinational projects.

## Leadership Training

According to the PMI (2013a) companies should have the following competencies to manage complex projects:

- A culture of project and program management with effective involvement of sponsors;
- Ability to teach and develop talents with a focus on promoting leadership skills;
- Ability to communicate effectively with all stakeholders groups.

To that end, leadership (project managers, PMO managers and others) should have leadership, technical and strategic capabilities, the latter related to the organization's business.

Leadership training should take into account that modern project management demands far more business objectives than technical objectives, as opposed to traditional project management. Kerzner (2006) makes the following considerations: in traditional project management there were 75% technical goals and the remaining business ones. Currently, in modern project management 90% of the objectives are business and the remaining technical ones.

In this context, leadership abilities are based on three factors: previous experience in project management, natural aptitude for leadership, and formal education in project management. It is natural that organizations need to empower potential leaders to make them real ones, since professionals with this profile are not usually available in the market or within the organization. Before the development actions of the leaders it is necessary to analyze the gaps between the various types of resources available and the resources needed for the complex projects, as well as an analysis of the gaps between the characteristics of the existing project managers and the characteristics of leadership demands for these projects.

One question raised by Kerzner (2006) was what is the point of balance between qualitative and quantitative trainings, whether for leadership or for the project team as a whole. Qualitative training, according to Kerzner, covers issues such as conflict management, motivation, leadership, team building and time management among others. They are training of behavioral inclination. The quantitative training includes planning, programming, cost control, software use and others.

The balance between the two types of training is conditioned by the company's culture, the specific characteristics of the project in question, and the characteristics of the team (and leadership) mobilized in the project in question. Determining this balance is one of the decisions made by the manager responsible for the training when the training is internal. If the training is external, it is likely that the company has limited management on this break-even point.

## Team Building

The quality of the actions of training (for the team) and support of the top management are two fundamental factors for achieving maturity, and following the excellence, in project management.

Training actions are an important knowledge management tool in that, when well executed, makes the necessary knowledge reaches those who need it for the correct development of the project.

All levels of a project team must be trained to be successful in managing this project. The objectives of the training actions are: to improve the knowledge and skills of the team in line with the demands of the project; promote teamwork and create a culture (cooperative) in project management.

In addition to training, capacity building: team building activities, codes of conduct and other ground rules, recognition programs, and tools for personal assessment.

We highlight here that one of the functions of PMO, proposed by the PMBOK (PMI, 2013c) is to train, guide and supervise. Thus, the PMO is an important agent for the qualification of the project team.

Training can be internal or external. Typically, large companies, when they choose to use an external training source, tailor the training program to their needs. Smaller companies usually opt for outside training with a more generic program.

The execution of the training activities usually goes through phases of identification of the needs, selection of the people to be trained, planning and conducting of the training and evaluation of the cost/benefit of the investment. Like many other management actions this sequence bears some resemblance to the PDCA cycle (plan, do, check, and act) disseminated by William Edward Deming.

## Knowledge Management and Ambiguities

A relevant consideration about qualitative training is that they can be used as a means to reduce ambiguities between the organization and stakeholder, primarily internal.

For example, training in codes of conduct, anti-corruption actions or the importance of safety and health conditions can give a clear indication of the type of relationship that the organization intends to establish with its employees and the communities where the project is located.

Note that when the company and its contractors employ local workers, the training given to them as employees (internal stakeholders) is reflected in the community (external stakeholders) of which they are a part.

Just as the boundaries between culture, structure, knowledge management and communication, the boundaries between internal and external stakeholders may not be clear and overlap.

## Knowledge Management and Stakeholders

We have already mentioned that third-party use is a reality in project management, that is, proprietary organizations are increasingly dependent on third parties as projects become more complex and multinational.

This reality has repercussions on how the organization manages its knowledge regarding project management, as discussed here.

By outsourcing and reducing the engineering functions of the proprietary organization, which are simultaneous and interrelated processes, these organizations have lost some of their skills (for project development).

One way to minimize these losses and their effects is to institute a repository of the proprietary organization for project knowledge.

This repository is the PMO, since some of its functions, as proposed by the PMBOK (PMI, 2013c), are:

- Identify and develop project management methodology, best practices and standards;
- Train, guide and supervise;
- Monitor compliance with project management standards, policies, procedures and models through project audits;
- Develop and manage project policies, procedures, models, and other shared documentation (organizational process resources).

This relationship, which is part of the relationship originally proposed in the PMBOK, highlights the activities that are directly related to the creation and maintenance of a collection of experiences and knowledge of the company for project management.

Although not mentioned explicitly in this relationship, it is obvious that one of the functions of the PMO, which supports the other functions listed above, is to organize the entire collection of lessons learned from the proprietary organization throughout its history as a project developer.

Another aspect to be considered, when there is participation of third parties, are contractual clauses that guide the exchange of knowledge between the owner organization and its service providers during the execution of the contracts. This establishes better conditions for the collection of knowledge managed by the PMO to grow consistently and remain relevant to its organization.

The PMO's role in knowledge management with stakeholders is to ensure that the events described by Prado (2015) will not happen:

*An inert set of standards in a binder, methodology or software installed on the computer network, but not used [...] They do not constitute a set of good practices because they were not continuously used by the main involved neither improved in the difficulties of the day by day work. There is with the maturation of practices that good practices emerge and we have good project governance.*

At the end, if the organization chooses not to institute a PMO, some area within its structure should be officially responsible for being the repository of knowledge in project management.

## Knowledge Management and Multinational Projects

Distance training, using the diversity of currently available communication technologies, is an important alternative for multinational projects, where staff is dispersed throughout the world.

These same technologies are an important communication tool in our globalized world, as we will see in the next topic.

# COMMUNICATION

The organization's ability to communicate with stakeholders has great influence on project development.

*Effective communication creates a bridge between diverse stakeholders who may have different cultural and organizational backgrounds, different levels of expertise, and different perspectives and interests, which impact or have an influence upon the project execution or outcome (PMI, 2013c).*

An example of this is the need to communicate the strategies associated with a project, as well as its goals, clearly since the beginning, as proposed by OPM3 (PMI, 2013b):

*Effective management organizations have a clear, well-communicated strategy and understand how each project supports it [...] At the beginning of a project, senior management communicates to the project team and provides answers to these important questions: What are the organization's long- and short-term operational goals and budgets? How does the project fit into or support these?*

The basic communication model has the components listed in Table 3.

Note that the model above is a simplification of a multi-dimensional process. Communication can be internal or external, formal or informal, vertical or horizontal, official or unofficial and can be written, oral, verbal or

*Table 3. Basic elements of the communication process*

| Component | Description |
|---|---|
| Sender | Who sends the message. The organization, for example. |
| Receiver | Who gets the message. The community where the project is located (external stakeholder), for example. |
| Transmitted message | It is the sent message itself, by the sender. |
| Message acknowledged | It is the acknowledgment that the message has been received by the receiver. |

Source: Author based on PMI (2013b)

non-verbal. Currently the various Social Networks available (on Smartphone, for example) bring even more dimensions to this process.

## Communication and Ambiguities

Returning to the model in Table 3, both the transmitted message and the message acknowledge are subject to noise. The noise will make it difficult for the receiver to have a correct understanding of the original message. Other noise, on the other hand, will make it difficult to understand correctly the acknowledgment of the message emitted by the receiver. Thus the company that has a good communication capability has to have the ability to minimize the noise that is caused internally as well as has to have the ability to deal with the noises that are generated externally.

Communication noises can enhance the ambiguities typical of complex projects. On the other hand, companies with good communication skills (and ability to deal with noises) can cope better and even reduce ambiguities.

For example: internally the communication should be used to disseminate to the whole team the project objectives, ensuring that they are known and understood.

## Communication and Stakeholders

Complex projects, that is, with multiple stakeholders, demand from the company a better communication capacity.

Peixto, Silva Neto and Mazzuchelli (2010) relate three weaknesses in the relationship with stakeholders, arising from the communication management process. Although the original study focuses specifically on mining companies in Brazil, the considerations are valid for complex projects in general:

- Start the Communication Management process, identifying stakeholders late.
- Disregard the existing relationship networks between stakeholders from different groups.
- Discontinue the relationship with stakeholders, within the Communication Management process.

Risk management and communication management processes must "go together" with regard to stakeholders. In addition to the need to start these two processes together, we can state that:

*In the project management language, the relationship with stakeholders is understood as" management "of the same, it is necessary that the translation of the same to the practice be perceived as a process of" dialogue (Peixoto, Silva Neto & Mazzuchelli, 2010).*

Note that the term "dialogue" with stakeholders is also used by Deloitte (2015) as the sixth topic of a list of the ten most important issues for mining today.

In the identification of stakeholders, companies should pay attention to the existing networks of stakeholders among different groups.

For example, mining projects are set up where available mineral and economically viable resources are available. Due to this, among other factors, the great majority of mining projects are installed in the interior of the states, obligatorily establishing a relationship with small municipalities.

By identifying the stakeholders, in any municipality, we can divide them into groups, such as: local entrepreneurs, political leaders and others. However, because it is a small township, it is very likely that one of the local entrepreneurs has a very strong, perhaps even kinship, relationship with a local political leader. Thus, if this local entrepreneur has a conflicting interest in the mining project, he or she may request support from the political leader, which may generate or increase the ambiguities in the relationship between the community and the project and ownership organization.

The Communication Management process should also not be discontinued. Throughout the development of a project, stakeholders can reposition themselves in relation to the project, generating more ambiguities and risks. The continuity of the Communication Management process, which implies the continuity of the dialogue with stakeholders, is essential for the company to perceive these repositioning as soon as possible, to analyze them and take the necessary measures to deal with it.

External stakeholders can use social networks to position themselves against those in favor of a project. Thus, the company must also be present in the social networks to position itself in the relevant issues of project and to follow the positioning of the stakeholders.

### Communication and Multinational Projects

A differentiated communication capacity in complex and multinational projects is fundamental for the proprietary organization to be able to deal with the cultural differences of the stakeholders.

Another important aspect for multinational projects is the communication technologies currently available, which contribute significantly to the development of the project by making the communication process feasible that were not possible for decades (or even years) ago.

## CONCLUSION

We opened this chapter by stating that our goal was to start a conversation about the culture and knowledge management aspects of mature organizations dealing with complex projects. In the sequence we added to the discussion two other aspects: structure and communication.

The choice of the expression "to initiate a conversation" was intentional, since we never had the intention to write something definitive about a subject so vast and complex. It is vast and complex because the interrelations between culture, structure, management of knowledge and communication are many and because we live in a world in what changes take place at an ever-increasing rate.

The proposition to "start a conversation" is in line with the character of maturity models, as these models provide guidelines and not process to be followed rigidly, as Persse and Persse (2001) state: "Operate within the guidelines, and you'll find lots of room to custom tailor and shape the program to the specific needs of your organization."

Regardless, we have advanced in our discussion we have established relationships between these four aspects and mature companies in project management, as well as with complex projects.

Table 4 aims to show these relationships in a summarized and schematic way. This framework allows the reader a quick return to the subject, whenever it is convenient or necessary. This table can also be used as a starting point for other insights on this subject.

The four aspects we address here also have interrelation with other aspects of project management, such as stakeholders and the risk management process. Thus, this chapter has fulfilled another function, that of creating a suitable context for discussing these two themes and others, which will be done in the subsequent chapters of the book.

*Table 4. Chapter Summary*

| | | Culture | Structure | Knowledge Management | Communication |
|---|---|---|---|---|---|
| Mature organizations | | A culture of cooperation and a focus on success. Project management is part of the organization's culture. | Culture prevails over the type of structure. Leader with leadership, technical and strategic skills. | Training of leaders and staff. PMO stores the knowledge of the company. | Diverse dimensions of communication. Social networks, especially on smartphones. |
| Complex projects | Ambiguities | Cooperative culture to deal with ambiguities. | Different places have different labor laws. | Qualitative training (code of conduct, for example) clarifies the company's positioning to stakeholders. | Minimize communication noises. |
| | Multiple stakeholders | Cooperative culture to deal with multiple internal stakeholders. | Participation of outsourced in the team. | PMO is a source of training for the entire team. | Relationship networks in communities. Dialogue with stakeholders. |
| | Multinational project | Cultural differences demand multicultural skills. | Virtual teams. | Distance or online training. | Communication technologies. |

Source: Author

## REFERENCES

Deloitee. (2015). *The top 10 issues mining companies will face in the coming year.* Retrieved from https://www2.deloitte.com/content/dam/Deloitte/global/Documents/Energy-and-Resources/gx-er-tracking-the-trends-2016.pdf

EY. (2016). *Top 10 business risks facing mining and metals, 2016–2017.* Retrieved from http://www.ey.com/gl/en/industries/mining---metals/business-risks-in-mining-and-metals

Kerzner, H. (2006). *Gestão de Projetos: As Melhores Práticas* [Project Management: Best Practices]. São Paulo, Brazil: Bookman Publisher.

Moura, A. A. (2013). *Gerenciamento de riscos associados a stakeholders em projetos de mineração (Monografia de especialização)* [Management of risks related to stakeholders in mining projects (Specialization Monograph)]. Belo Horizonte, Brazil: Pitágoras University.

Peixoto, A., Silva Neto, J. C., & Mazzuchelli, N. (2010). O Processo de Gerenciamento de Stakeholders em Projetos: Desvios da Prática Brasileira [The Process of Management of Stakeholders in Projects: Deviations from Brazilian Practice]. In *Anais, 5th Project Management Congress and 2nd Project Management Seminar in the 3rd Sector.* Belo Horizonte, Brazil: Academic Press.

Persse, J. R., & Persse, D. J. (2001). *Implementing the capability maturity model.* New York: John Wiley & Sons, Inc.

Prado, D. (2015). *Maturidade em gerenciamento de projetos* [Maturity in project management] (3rd ed.). Nova Lima, Brazil: FALCONI Publisher.

Project Management Institute. (2013a). Navigating Complexity. *Pulse of the profession in-depth report.* Retrieved from http://www.pmi.org/-/media/pmi/documents/public/pdf/learning/thought-leadership/pulse/navigating-complexity.pdf

Project Management Institute. (2013b). *Organization Project Management Maturity Model (OPM3)* (3rd ed.). Project Management Institute.

Project Management Institute. (2013c). *Project Management Body of Knowledge (PMBOK)* (5th ed.). Project Management Institute.

Silva Neto, J. C. (2011). *Avaliação de maturidade no gerenciamento de projetos em uma empresa de mineração em minas gerais* [Evaluation of maturity in the project management in a mining company in Minas Gerais State] (Masters dissertation). Fundação Mineira de Educação e Cultura, Graduate Program in Administration, Belo Horizonte, Brazil.

Section 2

# Theoretical Background:

## Understanding the Complex Projects and Maturity

# Chapter 4
# Introduction to Complex Projects


## ABSTRACT

*Complexity is an inherent feature of any project. It is used to classify and designate project managers. Explaining whether a project is complex, classifying this complexity and thus score for prioritization is often not a simple task. Each project will have its particularity, and this will be even more particular within each business sector. A $1 million project can be simple for a construction company, but of high complexity for an information technology company. This chapter will address common complexity concepts for different business sectors.*


## INTRODUCTION

Starting from the premise that, world development is becoming increasingly challenging and competitive for companies offering products and services. It is undeniable that inherent factors such as complexity of projects become the major hurdle of organizations. This means that the professionals involved in the development of the project will have the mission to overcome this challenge in an environment of globalization, dynamism and high competitiveness in which we live.

DOI: 10.4018/978-1-5225-5864-4.ch004

Thus, the need to develop new studies with the objective of verifying variables that are part of the project management process, such as the question of uncertainty, is emerging because the larger the company and the complexity of its projects, the greater the uncertainties involved, requiring a better management control and a more adequate connection of each project to its continuous operational processes (PMI, 2013).

But what determines this complexity? According to Vidal, Marle and Bocquet (2011 apud Souza, Novaski, Anholon, Besteiro, p.2, 2014), "project complexity is the property that a project has to become difficult to understand, predict and maintain under control its overall behavior, even when it has reasonably complete information about the project system. "

It can be considered that the complexity of a project is also linked to the difficulty of measuring the number of variables that directly and indirectly imply the development and control of a process. Although this does not seem to be the case, in some cases a project may present greater complexity given a particular business sector, considering its economic value equal to another project to be developed in a different follow-up.

Knowing and assigning a number of inherent variables to a project, and sorting them according to the degree of influence on complexity, is a laborious task for project management teams to develop efficiently and clearly. According to Whitty and Maylor, (2009 apud Souza, Novaski, Anholon, Besteiro, p.3, 2014), "the complexity in the project environment does not originate only from an individual structure of elements (external project stakeholders, task characteristics, and organizational complexity), and their interaction, but also with the dynamic effects of change generated by each of the interactions, and then with the interaction between the changes that were generated, consequently generating new changes in other parts of the system. "

In order to explain why projects are naturally complex, this chapter presents several concepts of complexity in general, based on the literature, and defines specific concepts of complexity that, when taken knowledge can become extremely important techniques and tools of use in Project Management of Engineering.

## DEVELOPMENT

Understanding the complexity of the project is important for successful project management, making the understanding directly connected with the difficulties faced, the decision making, and the achievement of the desired results for the project. There are many definitions of complexity. In order to address the relevance of complexity in engineering projects, this material focuses on the diversity of complexity concepts present in the literature, on the definition of complex projects, on the variables and uncertainties inherent in the projects as a whole.

Complexity of Latin (Complexus): it is defined by heterogeneous constituents inseparably associated, represented by events, actions, interactions, retroactions, determinations, accidents, that constitute our phenomenal world. But then complexity presents itself with disturbing traits of ambiguity, uncertainty [...]. This demonstrates how necessary it becomes for your understanding in project management.

Sinha, Thomson and Kumar (2001) argue that: "there is no single concept of complexity that can adequately aggregate the intuitive notion of what the word should mean." Projects have been described as complex systems that require management, not only because they involve technological issues, but because they involve large organizational factors, which are beyond the control of the project manager (Whitty and Maylor, 2009 apud Souza, Novaski, Anholon, Besteiro, p.2, 2014).

For Para Maximiano (2008, p. 07 apud Souza, Novaski, Anholon, Besteiro, p.5, 2014)the complexity is measured by the number of variables contained in a given project, so a complex project is one that has a large number of variables to be managed, such as: (a) multidisciplinary or diversity of professional profiles required to carry out a project; (b) physical distance between people or resources of the project; (c) the number of persons, organizations or facilities involved; (d) the diversity and volume of information to be processed; (e) duration; (f) conditions to be observed (risk and safety, for example).

As an important aspect, complex enterprises are characterized by multiplicity of variables. Few variables, although dependent on each other, may not be complex for projects. This multiplicity is associated with two other aspects: interaction between variables and unpredictability regarding their behavior (Brockmann, 2012 apud Andery & Veiga, p.2, 2013). This unpredictability may concern the role and responsibility of agents, the weight or level of influence of each variable, which may change over the course of project development, or the difficulty of defining and applying standards to construction and management processes. Therefore, a dynamic dependency between the variables is defined.

In this sense, the level of interaction and predictability in the behavior of variables could be an indicator of the degree of complexity. At the same time, in the understanding of BROCKMANN (2012 apud Andery & Veiga, p3, 2013), it would allow to distinguish between complex and complicated projects, being those in which there would be great interaction between variables, but with a high degree of predictability of their behavior. Thus, Figure 1 presents a basic outline of the influence of unpredictability on the level of complexity of a project.

In the same line of reasoning, Howell (2011) suggests that complex projects can be recognized when the relationship between cause and effect is not predictable, and therefore must be taken dynamically. Analyzing the point of view of each author mentioned above, it is possible to identify the basic variables that directly implicate the complexity of a project from the beginning. Thus, they can be represented within a macro context, according to table 1.

While many companies do deep and thorough analysis of a project's potential risks, it is not uncommon to find situations where managers and teams do not know how to act. This can happen both due to internal changes, as well as totally unexpected actions by one of the Stakeholders, as well as elements external to the project, for example, the variations in the market that impact the expectations about the product that was being developed.

*Figure 1. Influence of unpredictability on the complexity of a project*
*Source: Elaboration of the author, from the literature*

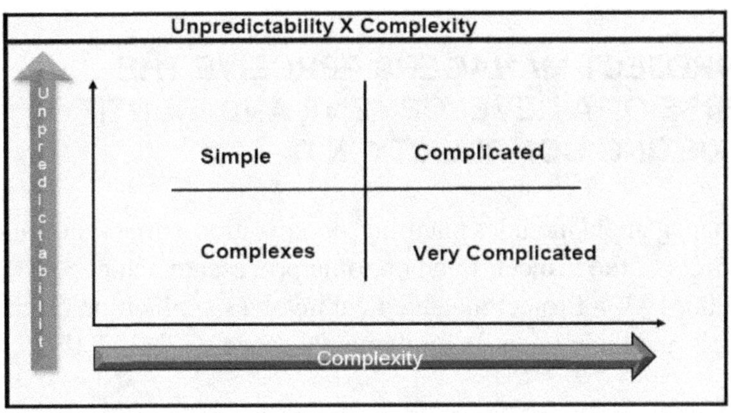

*Table 1. Basic attribute variables complexity in projects*

| Primary Variables | |
|---|---|
| **Variable** | **Concept** |
| People | The way people handle a project (whether managers, employees, customers or partners) is one of the great causes of complexity for a venture. This comes from the times of misaligned expectations, or lack of understanding of project objectives, ethical conflicts and instabilities created by members, and team changes. Events of this nature cause worsening communication problems between the parties involved in the process, which in the long run can eventually lead to the failure of the project. |
| Integration | It is considered a system, as the sum between different parts that produce maximized results that each of the parties could achieve by itself. It is easier to understand that the more elements involved in a project, the more complex it is to execute. These elements characterize not only the number of people involved in the project, which aggravates the factor (people), but also the interdependence of factors such as other internal or external projects and programs of the organization.<br>Integration can vary indefinitely depending on the branch of activity of the company, for example, in the companies that develop IT projects, the number of functionalities of a system (or its integration points like others) also contribute significantly to the increase of complexity as a whole. |
| Uncertainty | The basic concept of uncertainties in the project environment can be defined as the lack of clarity of project requirements and objectives, technical and project management knowledge, experience, technological novelty, information availability and reliability, organizational efficiency and tasks the teams. |

Source: Elaboration of the author, from the literature

The uncertainty of the factors are what make the lives of managers and teams more difficult, because working on projects where it is not possible to control a lot can be a real mess. However, it can be considered that the control of uncertainty in the initial phase of the project and throughout its development, is determinant for the success of the objectives defined in the scope. By demonstrating the impact of the variable uncertainty on the complexity of the project, figure 2 presents an illustrative schematic in the development of the process.

## HOW PROJECT MANAGERS PERCEIVE THE UNIVERSE OF A DEVELOPMENT, AND IDENTIFY THE EXISTING COMPLEXITY IN IT

Project management includes planning, organization, supervision and control of all aspects of the project, in an ongoing process, to achieve its objectives (ISO 10006, 1997). Project management involves application of knowledge, skills, tools and techniques as fundamental aspects (PMI, 2008).

*Figure 2. Uncertainty X Complexity*
*Source: Elaboration of the author, from the literature*

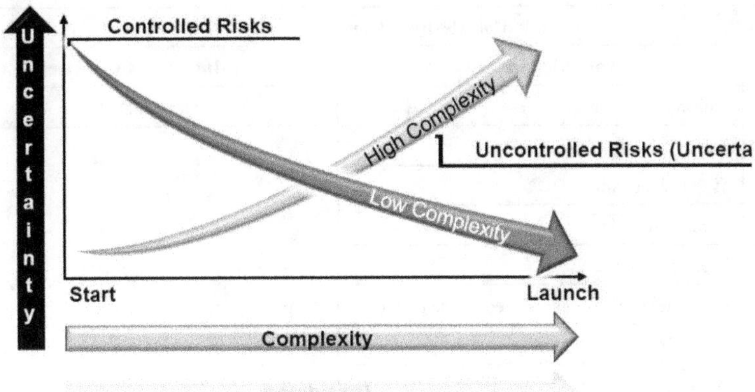


In this way, expressing each of the three basic variables mentioned above, and grouping particular variables within the three defined groups (People, Integration and Uncertainty), from the point of view of Project Management, the field of vision can be expanded to the influences that a project can suffer, according to Table 2.

In order to understand how complexity is inserted in the context of Project Management, it is necessary to define the concept of Project or Enterprise. Engineering project is the enterprise (business) whether industrial, civil, commercial or other, that should happen within an established period, negotiated cost, qualities imposed and scope agreed between Contractor and Contracted. Within an engineering project, there are decisive phases for its structuring, each phase being a particularity of the project necessary for its programming and organization. The set of phases, in most cases is constituted by the following basic processes: Engineering, Supplies, Construction and Tests.

According to The PMBOK Guide 5th Edition of Project Management of Knowledge (PMI, 2013) "planned lifecycles (also known as fully planned lifecycles) are those in which the project scope, as well as the time and costs required to deliver such scope are determined as early as possible in the project life cycle". As shown in Figure 3, the PMBOK Guide 5 defines that, "these projects progress through a series of sequential or overlapping phases, where each phase usually focuses on a subset of project activities and project management processes. The work performed at each stage is usually

*Table 2. Attribute variables complexity typology / characteristic influence in project management*

| Complexity Attribute Variables | | |
|---|---|---|
| **Number** | **Variable Typology** | **Influence Characteristic** |
| 1 | Project team size | People |
| 2 | Relative design size | Uncertainty |
| 3 | Type of innovation | |
| 4 | Organization size | |
| 5 | Type of project | |
| 6 | Impact level of changes caused by the project | |
| 7 | Type of project structure | |
| 8 | Pressure by deadlines | |
| 9 | Number of organizations involved in the project | Integration |
| 10 | Number of departments of the organization involved in the project | |
| 11 | Number of project stakeholders | |
| 12 | Dependency level of other organization projects | |
| 13 | Geographical location of stakeholders | |
| 14 | Location of project team members | |

Source: Elaboration of the author, from the literature

*Figure 3. Example of overlapping phases and engineering phase*
*Source: Elaboration of the author, from the studied literature (Guide PMBOK 5º edition Project Management of Knowledge, PMI, 2013).*

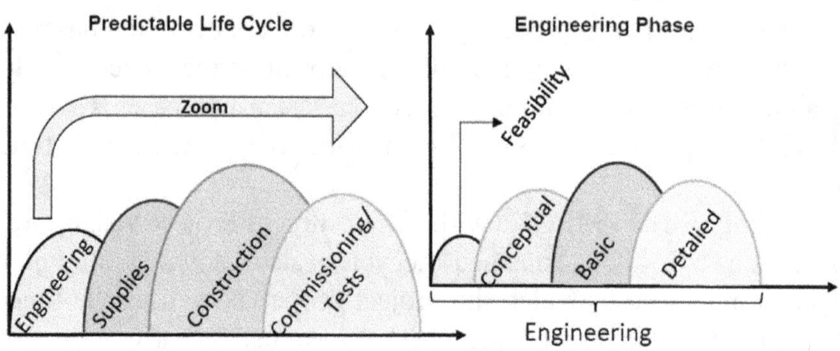

of a different character from the work of the previous and subsequent phases and thus the required training and skills of the project team may vary from stage to stage. "

In projects with more than one phase, there may be different relationships between the individual phases (overlapping, sequential, parallel). Considerations such as the level of control required, effectiveness and degree of uncertainty determine the ideal relationship between phases. Still according to figure 2, it is possible to expand the engineering phase and define the processes and variables that infer in its development.

Thus, it is possible to realize that, contained in the engineering phase, there are fundamental subphases constituting the planning process of the main phase. Being that the complexity becomes present from the evaluation of the viability of a project. Table 3 presents examples of the specific variables inserted in the viability phase.

In Project Management there are also other concepts that define the complexity of an enterprise can be understood within four classes of complexity.

- **Structural Complexity:** First type of complexity and the most cited. It is related to a large number of distinct and interdependent elements. Most of the articles researched by authors define structural complexity based on three attributes: size (or number), variety and interdependence. It includes the organizational and technical dimensions proposed by Baccarini (1996, apud Lafetá, Barros, & Leal, p 4, 2016);

*Table 3. Specific variables that affect the viability of a project (uncertainties)*

| **Variables That Affect the Viability of a Project** | |
|---|---|
| Technical Study | Each project has its own different technical variable. The technical approach is the first step of the project feasibility study. It is from this that the other steps will be followed. It will be up to the body of engineers to tell if the project scope can be executed. |
| Financial Study | The financial feasibility study will draw on a lot of technical study information, such as the size of the team, the required professionals, the completion period x investment, the estimated total cost of material and man-hours.<br>In summary, financial viability is the calculated cost of the project in more than one condition of term and volume of investment. In order for the project to be feasible, it is necessary to finance and need an investment capacity of those involved, there is no enterprise for the height of at least one thing projection. |
| Economic Study | It is assumed that every work has a purpose and that this purpose is related to profit. The ultimate goal is whether the project is economically viable, whether the investment will be recovered through the operation, how long it will take, and what the potential for revenue and profits will be for subsequent years. |

Source: Elaboration of the author, from the literature

- **Dynamic Complexity:** Refers to changes in projects, such as changes in specifications (or changes in objectives due to ambiguity - concept related to "uncertainty"), project management team, suppliers or environmental context. (Lafetá, Barros, & Leal, 2016) cites as an example the dynamic structural complexity that would result from a change of scope, the dynamic uncertainty arising from the emergence of a new technology, or the rhythm changes from the imposition of a new deadline for completion.
- **Complexity of Rhythm:** The rhythm dimension presents itself as an important type of complexity. It emerges as urgency and criticality in the deadline for meeting goals that normally require different structures and greater managerial attention. Williams (1999) emphasizes the need for simultaneous engineering to meet tight project deadlines, which leads to a closer interdependence between the elements of the system, which therefore intensifies its structural complexity. Unlike other types of complexity, rhythm is not an abstract construction with several indicators and aspects. It refers essentially to the rate at which projects are (or should be) delivered.
- **Sociopolitical Complexity:** This type of complexity is directly related to the participation of human actors in the processes that make up the projects, which have potentially conflicting interests and difficult personalities. It adds the diversity of mental models and values of the decision makers involved.

Dunovic, Radujkovic and Skreb (2014 apud Lafetá, Barros, & Leal, p.2, 2016) emphasize that project managers often recognize the complexity too late to control the situation. Therefore, they should be aware of the complexity of the project from the outset in order to develop appropriate strategies and designate the appropriate team members.

## CONCLUSION

With the knowledge of several concepts of complexity defined by different authors, this chapter demonstrated a series of factors classified as variables, which influence the complexity of each project. These factors interact with each other, enhancing the complex universe of ventures. What is not possible, is to determine a generic standard of complexity for all projects, since, according to the text, each project has its own particularities and its influential variables. But it became clear that some basic variables like people, integration and uncertainty can be inserted into any type of project.

The variables of complexity of the projects must be known and worked by the managers of the organization and of the projects, allowing a specialized treatment to these projects. Planning teams must be able to handle the complexities of the projects, developing specific techniques and processes. Factors that add high levels of complexity should be analyzed in detail in their causes. Likewise, factors that minimize complexity should be encouraged and better developed.

Another observed behavior is that the complexities of the projects may vary over time as the circumstances and characteristics of the projects change. Management actions and improvement plans can also reduce the impact of the complexity of the projects. As the projects are composed of phases that are placed on top of one another, giving continuity to the progress of the process, it is necessary for the variables of the complexity attribute of each phase to be identified and dealt with before proceeding to the next stage of the project, ensuring that risks are controlled throughout the development of the business until its launch. In this way the maturity acquired by the organization over time also contributes to minimizing the effects of complexity on a whole.

## REFERENCES

Andery, P. R. P., & Veiga, A. C. R. (2013). *Considerações sobre o gerenciamento de projetos complexos: o caso de exposições museográficas*. Disponível em: http://www.fumec.br/revistas/construindo/article/view/2104

Lafetá, F. G., Barros, F. D. O., & Leal, P. D. O. C. D. (2016). *Projetos Complexos: Estudo De Caso Sobre A Complexidade Dos Projetos De Engenharia De Telecomunicações Em Uma Empresa Do Setor De Óleo E Gás*. Disponível em: http://www.revistagep.org/ojs/index.php/gep/article/view/382

Pinto, J. S., Novaski, O., Anholon, R., & Besteiro, E. N. C. (2014). *Proposta De Método De Mensuração Da Complexidade Em Projetos*. Disponível em: http://www.revistagep.org/ojs/index.php/gep/article/view/280

Project Management Institute – Library of PMI Global Standards. (2018). Disponível em: http://www.pmi.org/PMBOK-Guide-and-Standards/Standards-Library-of-PMI-Global-Standards.aspx

Souza Pinto, J. D., Novaski, O., Anholon, R., & Besteiro, E. N. C. (n.d.). *Variáveis Dos Atributos Complexidade E Incerteza Em Projetos: Um Estudo Exploratório*. Disponível em: http://www.inovarse.org/sites/default/files/T14_0427.pdf

# Chapter 5
# Types of Complex Engineering Projects


## ABSTRACT

*Works, constructions, and fabrications are examples of projects that in themselves are complex, especially for inexperienced project managers or other areas. The complexity classification of engineering projects is important to determine the necessary efforts for the management of this enterprise and also to analyze threats and opportunities. This chapter looks at how this categorization is done and the best practices in engineering and construction projects. At the end, it presents a proposal to classify the complexity of enterprises in energy, civil construction, mining, and other related business sectors. This is a conceptual and applied reading in engineering and construction projects.*


## INTRODUCTION

Great achievements in history such as the construction of the wall of China, the Egyptian pyramids, have similarities with great modern achievements such as the construction of the Panama Canal, the Itaipu Hydroelectric Power Plant and the Eurotunnel. This similarity also extends to the modern companies of computational systems developments that seek to implement software capable of solving complex problems of diverse areas in the daily life. Other current sectors that also have great achievements of projects, is the mining and civil construction with focus in ventures of Shopping Centers. Faced with this similarity of achievements that occurs in times, places and with such

DOI: 10.4018/978-1-5225-5864-4.ch005

diverse goals, is a success for each of the achievements of unique character. These successes require, since ancient times, a disciplinary application of techniques, tools, organizational systems and especially of people gifted with skills and knowledge for conflict management, and propose problem solving.

After analyzing this relationship between projects from a more distant past and contemporary ventures we can think of three sectors of great importance in the world economic senate, even if there are indifferences between developed and underdeveloped countries, which are subject to develop projects of very high levels of investment values and complexity. It can be said that among them are the energy sectors, with their large projects of hydroelectric power plants providing autonomy and independence in the generation of electric energy, the mining sectors that can represent internal indicators in the growth of a country's GDP, and the commerce sector with large enterprises of the construction of Shopping Centers generating jobs and developments in the retail trade.

Thus, deepened in each of the sectors mentioned above, it is possible to analyze what makes these projects complex. How can you measure the complexity inherent in these large enterprises? What issues should be raised before these projects to define their complexities, making it possible to maintain control of the goals and scope of the objectives defined in the scope, preserving the economic, political, environmental, and project success interests? What are the challenges of constructability and execution of projects of this nature?

The challenge posed by megaprojects is directly linked to its complexity. A complex system is different from a complicated system, since it has a type of dynamism that actively responds to what is happening around it, becoming unpredictable (Agostinho, 2003 apud Oliveira, Winter, Alvim, & Corrêa, p.1, 2014).

As with cases of complexity in hydroelectric projects, the Portal do Estadão de São Paulo published on November 3, 2017 a report that shows the paralysis of a project to expand hydroelectric power plants in the Amazon region of Brazil, due to its complexity socio-environmental. "The pressure on hydroelectric reservoirs tends to increase in the coming years, given the absence of new projects based on the construction of large dams and hydropower generation. After a decade of investments in large enterprises, such as Jirau, Santo Antônio, Belo Monte, Teles Pires and São Manuel, all installed in the Amazon region, the government virtually paralyzed the construction of new projects, due to the environmental complexity of these projects " (Portal do Estadão de São Paulo, 2017).

In mining projects, an example of the complexity involved is presented by Oliveira and Andery (2017), and "mining projects are usually complex, and often have performance problems associated with increasing the deadlines for entry into operation. In this sense, one of the major bottlenecks, considered as a critical path in the mining project schedules, is environmental licensing. Environmental licensing processes have become increasingly rigorous, with more demanding responsible bodies, which have required a significant period of time in the project schedules, often leading to considerable impacts on the implementation of the projects."

However, in construction projects and shopping malls, there are also examples of the influence of complexity on projects. According to Couto (2012), "based on daily work situations in these works, we verify that commonly used artifacts in collections to receive projects, such as periodic meetings in which dates are set for project delivery, suspension of payments and binding are not sufficient to meet deadlines and then a deadlock is created between the requirements of the entrepreneur, who needs a return on investment and to honor contractual commitments related to the inauguration date of the project that were signed with the tenants, the demands of the work and the capacity of production and delivery of the projects. Associated to this, we have a particularity in works of Shopping Centers that is the mix of the stores that is constantly changing, since the selection of the stores is of great importance for the quality and longevity of the enterprise. In this way, the commercial aspect overlaps in most cases with the technical aspects. We can say that Shopping is a "living organism" that is in constant change of mix and consequently of the layout of the stores and the enterprise as a whole, which turns out to be an additional element in the management of the deadlines / schedules ".

Based on this context of the complexity that covers projects of different areas important for the development of the economy sector, this chapter presents techniques and tools conceptualized by scholars, which point out the possible causes of complexity in Engineering and Construction projects in general, and proposes ways to merge the various techniques by pointing out the complex elements in the Engineering, Supplies, Construction, and Commissioning phases.

# DEVELOPMENT

## Classification of Complexity of Projects

To understand the fundamental means by which organizations can classify their projects, we can examine the classical contingency of the theory of innovation. This theory states that different conditions may require different organizational environments, and that organizational effectiveness is contingent on the amount of adequacy between structural and environmental variables. But how do classical contingency arguments remain in the dynamic, temporary, and unstable world of projects? Unlike companies, projects are temporary organizations. They are limited in time, usually part of a larger organization, and perform mostly new tasks that have not been done previously (Lynn CrawfORD, Brian Hobbs and Rodney J. Turner), Project Categorization Research Report (NEWTON SQUARE, PA: Project Management Institute, Research Development, 2004).

Therefore, in search of greater distinctions between projects, several observations can be made. First, the fundamental nature of projects as tasks that were never done, of course, leads us to consider project uncertainty as a major dimension for selecting projects. Second, some projects may be more complex than others, and so the complexity of the task and organization is another clear candidate for distinction. Notably, the combination of uncertainty and complexity has also often been mentioned as a basis for distinction. Finally, since each project has a time limit, we can also consider the time constraint as a basis for project differentiation, and can classify it according to the pace of achievement of the stipulated goals. Figure 1 depicts the interaction between three variables considered in the primary process of classifying a project.

Examining in more detail each of the variables involved in the dimensions of a project, we can try to see how they affect the in the classification of the complexity of the ventures.

- **Uncertainty**. Uncertainty means how much we do not know at the beginning of the project. Different projects present, at the beginning, different levels of uncertainty. Project uncertainties can be external or internal, depending on the environment or specific task and the ability to carry it out.
- **Complexity**. It depends on the complexity of the product and the task - more specifically, the product structure and functionality, as well as the

*Figure 1. Basic variables for project classification*
*Source: Prepared by the author himself from the referenced literature*

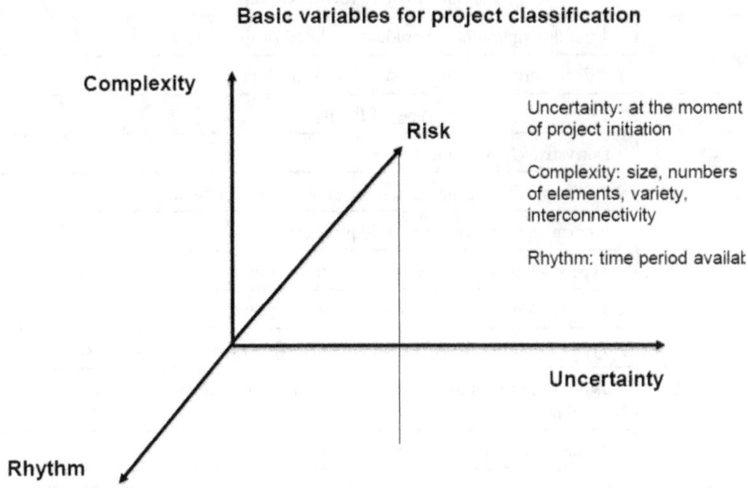


number and variety of elements, the subtasks, and the interconnections between them.

- **Rhythm**. The third dimension involves the urgency and criticality of time goals. The pace depends on the time available allocated for the completion of the project and the degree of urgency.

Another artifice that also contributes to determine the classification of projects is the application of a questionnaire raising information about strategic points of the project. It is necessary to clarify all the objectives of the project and the parties involved in it from the scope, Table 1 presents relevant questions to be observed in the idealization of the enterprise.

This questionnaire should be completed with the following questions:

- **Technology Uncertainty:** Low - Medium - High - Super High, required for the project.
- **Complexity:** Assembly - System - Matrix, Dependent for execution.
- **Rhythm:** Regular - Fast - Critical time to achieve goals.
- **Business Goals:** Operational - Strategic
- **Client**:External – Internal
- **Strategic Goal:** 1 Extension - 2 Strategic - 3 Problem Solving - 4 Maintenance - 5 Utility - 6 Research and Development.

*Table 1. Project classification*

| Primary Project Identification | |
|---|---|
| Product description | A brief description of the product produced by the project: |
| | A brief description of the scope of work in the project: |
| **Type Of Project** | |
| 20X Product News | 1 Derivative (Improvement). |
| | 2 Platform (A new generation in an existing product line). |
| | 3 Innovation (A new-to-the-world product). |
| 30X Uncertainty Technological | 1 Type A - Low-Technology (No new technology). |
| | 2 Type B - Medium-Technology (Some new technology). |
| | 3 Type C - High-tech (All or almost all new but existing technologies). |
| | 4 Type D - Super-high technology (Project will use technologies not existing at the beginning). |
| 40X Complexity (System Scope) | 1 Assembly (A subsystem - performing a single function). |
| | 2 System (A collection of subsystems - performing multiple functions). |
| | 3 System (System of Systems - A widely dispersed collection of systems serving a common mission). |
| 50X Rhythm | 1 Regular (Delays are not critical). |
| | 2 Fast / Competitive (Time to market is a competitive advantage). |
| | 3 Critical time (Time to completion is critical to success, window of opportunity). |
| | 4 Blitz (Crisis Projects). |
| 60X Business Goals | 1 Operational (Extension of an existing business). |
| | 2 Strategic (Creation of a new business). |
| 70X Customer | 1 External (Contracts or external consumers). |
| | 2 Internal (Internal users or another department). |
| 80X Strategic Goal | 1 Extension (Improving, improving an existing product). |
| | 2 Strategic (Main - creation of new strategic positions in business through new products or markets). |
| | 3 Troubleshooting (Acquires or develops new technology or new capacity). |
| | 4 Maintenance (Routine maintenance, problem solving). |
| | Utility (Keep lights on - purchase and install new equipment or software, implement new methods or new processes, reorganization, reengineering). |
| | 6 Research and Development (Study - exploration of future ideas, no product specific in mind). |

| Deadline for Execution | | |
|---|---|---|
| Start Date Project (Month year) | Project Duration (Months) | Budget ($) |

Source: Prepared by the author himself from the referenced literature

After identifying the project making the observations described in table 1, it is also possible to make a representative scheme according to figure 2, which represents the level of complexity relating to technological variables as a function of the novelties proposed by the project.

In order to classify the complexity of an enterprise according to technology, the following aspects are taken into account.

## TECHNOLOGY

- **Low**: the company already dominates this technology, well-established and mature technology.
- **Average**: Only product details present technology (existing) but not dominated by the company. It is not a critical part of the product.

*Figure 2. Influence of technology, innovation and pace in the complexity of projects*
*Source: Prepared by the author himself from the referenced literature*

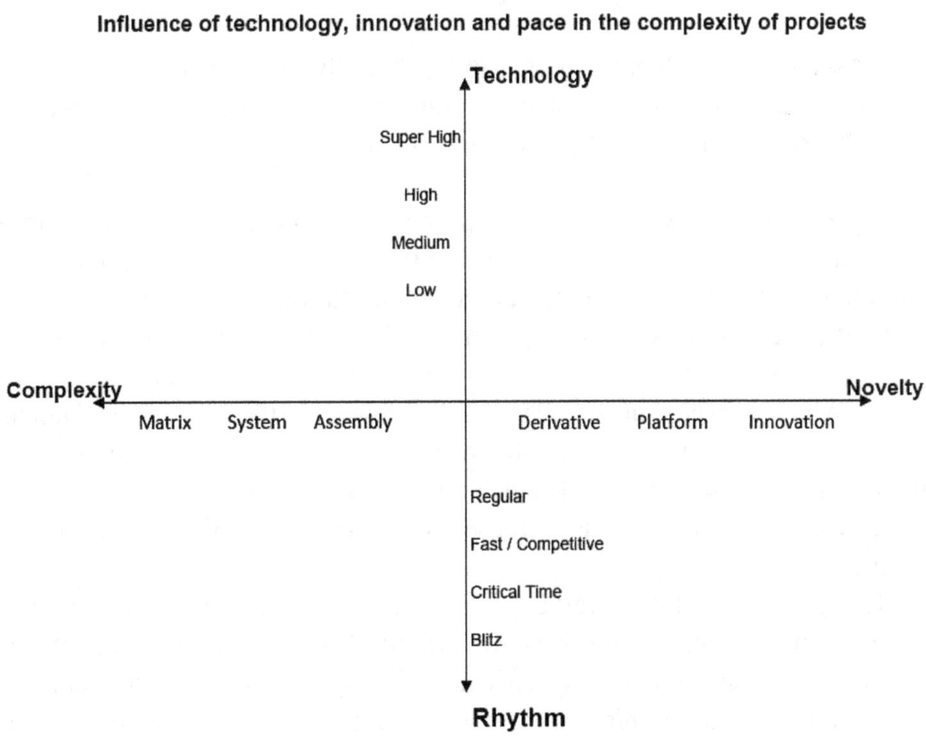

- **High**: uses new, newly developed but existing technologies. First use of new technology. Consider the company's benchmark: technology for the company.
- **Super high**: technology does not exist, will be developed during the project

## NEWS

- **Derivative:** new version of an existing product.
- **Platform:** new line or family of products.
- **Innovation:** new type of product never before thought by the consumer

And consequently the measurement of the complexity according to the parameters of production, and parties involved.

## COMPLEXITY

- **Assembly:** product that exists independently of the rest.
- **System:** product that involves several subsystems, all are interconnected.
- **Matrix:** union of multiple systems, large scale

Figure 3 shows an example of a shopping center project comparing it to a conventional house project. Considering that a large enterprise, such as the shopping center projects, demands the classification of countless variables that influence the level of complexity of the project. In this example the variables inherent to the two projects appear in four specific points.

It can be observed that, a mall venture compared to the conventional house project, demands greater technology, innovation and pace in meeting the goals, which consequently influence the level of complexity of the project. In this case, the complexity is increased from the integration of the following projects:

Topography; Architectural; Lighting Technician; Interiors; Landscaping; Visual communication; Road system; Structural, may be metallic, concrete or mixed; Building installations - electrical, telephony, automation, hydro sanitary, fire prevention and fire fighting; Air conditioning, mechanical ventilation and exhaust; Given these aspects, the complexity is represented by the high cost

*Figure 3. Example of the complexity level of a shopping mall project due to technology, innovation and rhythm*
*Source: Prepared by the author himself from the referenced literature*

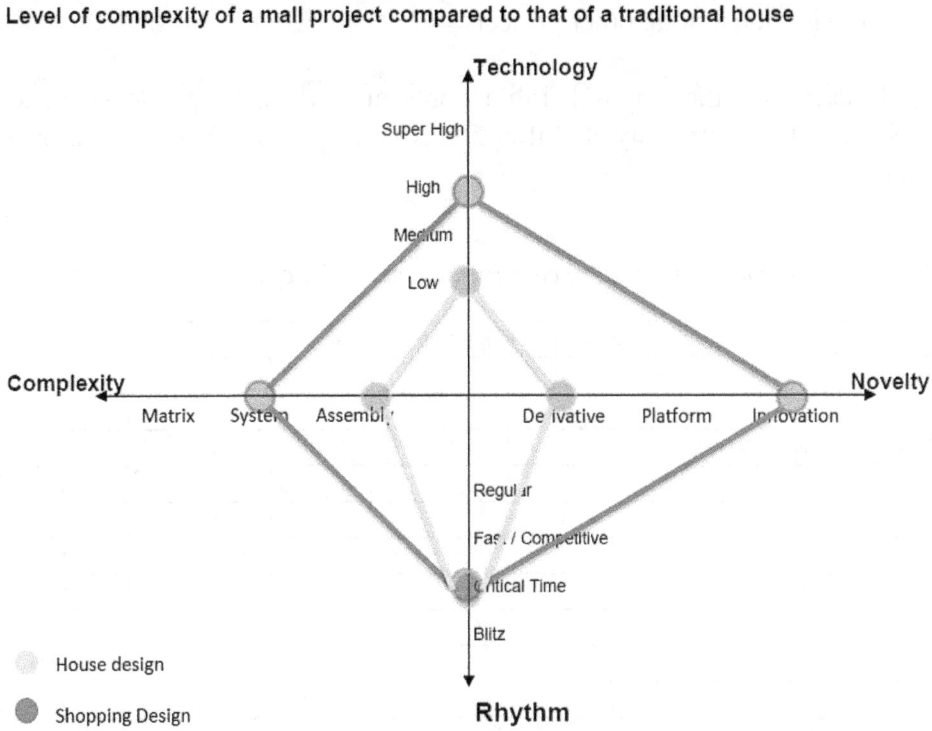


of investment in the project, time of implementation and execution, number of servers involved in the process and integration between projects.

## Proposal for Classification of the Level of Complexity of Small And Medium-Sized Porte Projects

The dimensioning of the complexity of an enterprise encompasses a huge context around the project. From the initial idea, feasibility, technology required, skilled workforce and bureaucratic documentation for project licensing. In addition to the previously cited ways of measuring the level of complexity of a given project, it should be determined through the following analysis.

- Cost of investment to carry out the project
- Time of implementation or realization of the project

- Amount of employees required for complete execution of the construction process
- Participation of external resources
- Integration among other projects

To exemplify the proposal, Table 2 presents a parameter model of values in scale of the complexity classification according to the criteria mentioned above.

*Table 2. Example of proposal to classify the complexity of projects*

| **Parameters for Project Measurement[PUB-71]** | | |
|---|---|---|
| Complexity | Size | Weight |
| Low Complexity | Little | 1 |
| Average Complexity | Medium | 2 |
| High Complexity | Large | 3 |
| Cost | Size | Weight |
| Up to $ 80.000,00 | Little | 1 |
| From $ 80.000,001 to $ 650.000,00 | Medium | 2 |
| Above $ 650.000,00 | Large | 3 |
| Time | Size | Weight |
| Up to 30 day | Little | 1 |
| From 1 to 6 months | Medium | 2 |
| Above 6 months | Large | 3 |
| Number of servers | Size | Weight |
| From 1 to 2 | Little | 1 |
| Between 3 and 6 | Medium | 2 |
| Above 6 | Large | 3 |
| External resource share | Size | Weight |
| No | Little | 1 |
| Yes | Large | 2 |
| Interconnection between projects | Size | Weight |
| Does not have integration with other projects | Little | 1 |
| Has integration with a project | Medium | 2 |
| Has integration with more than one project | Large | 3 |
| Weight | | |
| Little | 1 | |
| Medium | 1,5 | |
| Large | 2 | |

Source: Prepared by the author himself from the referenced literature

The filling of table 2 is performed considering the weight factor with a score of (1, 2, 3) to classify the complexity and the processes of the project.

Being:

## Size of Projects X Complexity Weight

- Small project - Complexity - Weight (1);
- Average design - Complexity - Weight (2);
- Large Project - Complexity - Weight (3)

Thus, considering the sector of activities in which the project will be proposed, it should be assigned the grade (Weight) for each project process, as follows:

## Investment Value

- Below ($): Small - (1)
- From ($) to ($): Medium - (2)
- Above ($):
- Up To (x) days: Small - (1)
- Between (x) to (x) months: Medium - (2)
- Over (x)
- High - (3)

## Project Implementation Time

- months: Large - (3)

## Number of Servers Involved in the Project

- Up To (x) days: Small - (1)
- Between (x) to (x) months: Medium - (2)
- Over (x) months: Large - (3)

## There Is Participation in External Resources

- [NO] Small - (1)
- [YES] Large - (3)

## Interconnection Between Projects

- No integration with other projects: Small - (1)
- Has integration with a project: Medium - (2)
- Has integration with more than one project: Large - (3)

Finally, an arithmetic mean of the total weight assigned to the project processes is classified, by classifying the total complexity of the project, being between the values of (1 to 3).

To consolidate another method of classification of complexity, Table 3 presents a model that classifies (criteria of complexity x strategic criteria) of enterprise, considering the following criteria: the constructability, budget, duration, technical aspects, legal demand and impacts on the strategy from the company. To classify the weight of each criterion, the table presents a score of the percentage of relevance in the project. With this it is possible to point out in the levels of complexity of the project, indicate parameters of investment values, and demand for specific qualification of labor, acquisitions and market experiences.

Following the example of table 3, filling it out requires that the following data be assessed by assessing the availability or demand of the company for each project criterion:

Levels of complexity are ranked in increasing order (1; 2; 3; 4), varying according to the way the project fits the criteria.

For each criterion, a grade (Weight) is assigned varying in a percentage related to the influence of each criterion in the project.

## Criteria

- **Constructability**: Experience that the company has or demand, on the type of project to be developed. Weight - (40%)
- **Budget**: Value of investment for the enterprise. Weight - (15%)
- **Time**: Rate of execution of the goals for completion of the enterprise. Weight - (15%)
- **Technical Aspects**: Defined by the qualification of the technical team, with adequate level of knowledge, or need greater knowledge and market experience. Weight - (30%)
- **Acquisition**: Involves new acquisitions of equipment, services, bidding or direct contracting.

*Table 3. Example of project classification - complexity criteria x strategic criteria*

| CLASSIFICATION OF PROJECTS - COMPLEXITY CRITERIA X STRATEGIC CRITERIA | | | | | |
|---|---|---|---|---|---|
| COMPLEXITY CRITERIA | | | | | |
| CRITERION | Level | | | | Weight |
| | 4 | 3 | 2 | 1 | |
| Constructability | We have experience and a reasonable chain of approved suppliers. | We have experience and a small chain of approved suppliers. | We have experience but no chain of suppliers developed. | Some experience in this type of project and no chain of suppliers developed. | 40% |
| Budget | Above R$ 10 million | Between R $ 5 million and R $ 10 million | Between R $ 1 million and R $ 5 million | Below R $ 1 million | **15%** |
| Technical Aspects * | It involves application of new knowledge and the team has little or no experience in the subject | It involves application of knowledge already widespread in the market, but the team has no experience in the subject | It involves application of new knowledge and the team has experience in the subject | Involves application of knowledge already widespread in the market and the team already has experience in the subject | **15%** |
| Acquisition | | | Involves procurement of equipment, products and / or services with Bid | Involves purchases of equipment, products and / or services with Direct Contracting | |
| * In this criterion knowledge involves not only technology but also new methodologies, models, patterns and changes in work processes; besides new ones | | | | | |
| STRATEGIC CRITERIA | | | | | |
| CRITERION | Level | | | | Weight |
| | 4 | 3 | 2 | 1 | |
| Legal claim | Legal claim with a term of less than 1 year | Legal claim with term between 1 and 2 years | Legal claim with a term of more than 2 years | No legal action | 40% |
| Impact on COMPANY Strategy | Project has a lot to do with strategy. | Project has significant relationship with strategy. | Project has some relation with the strategy. | Project has little relation to strategy. | 60% |

Source: Prepared by the author himself from the referenced literature
Models to evaluate complexity, available at: http://www.pmkb.com.br

- **Legal Demand**: Implies in the time necessary to collect the necessary documentation for the licensing of the project. Weight - (40%)
- **Impact on Comp any Strategy**: Determines the level of impact on company strategies, caused by the project. Weight - (60%)

At the end, you can assign the final score by making a relation (COMPLEXITY CRITERIA X STRATEGIC CRITERIA), of the assigned scores from (1 to 4) relating according to Table 4.

## CONCLUSION

After studying this chapter, a new look at the understanding of complexity in engineering projects will be promoted. The concepts and techniques presented help in the identification of complex criteria of the projects, guiding a more concise and based decision making, allowing assuming lower risks inherent in the complexity of projects. It becomes feasible a simultaneity in the application of the proposed models, analyzing the best practice to be applied according to the proposal and identification of the project. In this way, we can link the various engineering niches involved in the project. Thus a greater scope will occur in the identification and classification of the existing levels of complexity of the enterprise, improving the strategic development.

Based on the references of this study, it is understood that the strategic tools to identify the complexity of projects can be applied to time management in projects of the most diverse works, since it seeks to achieve a high level of performance, reliability and value for the customer. Allied to this, the identification of the root causes the implantation of improvements of quick gain and the loss minimization are important tools for analysis of the process of elaboration and execution of the projects that can serve as guide for the definition of management actions.

*Table 4.*

| Complexidade | | | | |
|---|---|---|---|---|
| 4 | 4 | 8 | 12 | 16 |
| 3 | 3 | 6 | 9 | 12 |
| 2 | 2 | 4 | 6 | 8 |
| 1 | 1 | 2 | 3 | 4 |
| | 1 | 2 | 3 | 4 |
| | Alinhamento Estratégico | | | |

De 1 a 2,99 – Baixa Verde

De 3,00 a 7,99 – Média Laranja

De 8,00 a 16,00 – Alta - Vermelho

## REFERENCES


Bahia, F. B. (2009). *Análise de Critérios e Fatores de Sucesso em Projetos de Engenharia, Suprimentos e Construção (EPC) Offshore*. Disponível em: http://www.bdtd.ndc.uff.br/tde_arquivos/14/TDE-2010-02-05T101846Z-2368/Publico/Dissertacao%20Fabio%20Bahia.pdf

Borges, A., & Warth, A. (2017). *Grandes hidrelétricas estão fora de planos de expansão*. Disponível em: http://economia.estadao.com.br/noticias/geral,grandes-hidreletricas-estao-fora-de-planos-de-expansao,70002071622

Couto, M. L. (2012). *Gestão Do Tempo Em Projetos Para Obras De Shopping Centers No Cenário De Aquecimento E Expansão Do Setor Varejista Nacional*. Disponível em: http://acervodigital.ufpr.br/bitstream/handle/1884/39599/R%20-%20E%20-%20MARIANA%20LARA%20COUTO.pdf?sequence=1



Couto, M. L. (2012). *Gestão Do Tempo Em Projetos Para Obras De Shopping Centers No Cenário De Aquecimento E Expansão Do Setor Varejista Nacional*. Disponível em: http://acervodigital.ufpr.br/bitstream/handle/1884/39599/R%20-%20E%20-%20MARIANA%20LARA%20COUTO.pdf?sequence=1



Fontoura, J., & Carneiro, L. (2009). *Experiência no Gerenciamento da Engenharia em Empreendimentos em Regime EPC*. Disponível em: http://www.brasilengenharia.com/portal/images/stories/revistas/edicao591/Eng_Gerenciamento.pdf

Junior, C. A. V. (2017). *Classificando Para Gerenciar Melhor*. Disponível em: http://www.eq.ufrj.br/docentes/cavazjunior/diamante.pdf

Ney, G. M. A., & Fortes, M. Z. (2017). *Metodologia Para Processo Integrado De Comissionamento Em Unidades Industriais*. Disponível em: http://www.revistasg.uff.br/index.php/sg/article/viewFile/1189/676

Okabayashi, A., Bruno, M. A. C., & Sbragia, R. (2008). *Construction Management Excelence Of Major Hydroelectric Power Plants*. Disponível em: http://www.scielo.br/pdf/ram/v9n6/a02v9n6.pdf

Oliveira, C. D., Winter, A. R., Alvim, T. S. R., & Corrêa, D. (2014). *Gerenciamento de Obras vs. Construtibilidade vs. Complexidade*. Disponível em: http://pmkb.com.br/artigo/gerenciamento-de-obras-vs-construtibilidade-vs-complexidade/

Oliveira, L. P. S., Shimada, R. J., & Moraes, R. M. (2013). *Análise De Empreendimentos Do Tipo Shopping Center: Influência Do Mix De Lojas E Da Localização Nos Indicadores De Qualidade Do Empreendimento – Um Estudo De Caso*. Disponível em: https://www.eec.ufg.br/up/140/o/an%c3%81lise_de_empreendimentos_do_tipo_shopping_center__influ%c3%8ancia_do_mix_de_lojas_e_da_localiza%c3%87%c3%83o_nos_indicadores_de_qualidade_do_empreendimento_-.pdf

Oliveira, P. M., & Andery, P. R. P. (2017). *Processo De Licenciamento Ambiental No Projeto De Empreendimentos Industriais De Mineração*. Disponível em: https://www.revistas.usp.br/gestaodeprojetos/article/view/111519

# Chapter 6
# Maturity and Models Around Project Management


## ABSTRACT

*Initially, this chapter discusses what maturity is in project management, what maturity models are, and what their role is. In the second part, this chapter looks at four of the major maturity models currently in use, describing the structure and talking about their application. The purpose of this analysis is to support later discussions in this book. In a world where complex projects are increasingly frequent, project management must provide organizations with a set of managerial processes to survive in this scenario of fierce competition and scarcity of resources.*


## INTRODUCTION

In a world where complex projects are increasingly frequent, project management must provide organizations with a set of managerial processes to survive in this scenario of fierce competition and scarcity of resources.

Because they are complex, most projects require effective management throughout their life cycle. Thus, seeking an increase in the level of maturity in project management has been a frequent concern of organizations.

In this context we can say that the current project management is characterized by:

DOI: 10.4018/978-1-5225-5864-4.ch006

- Demand project management skills not only from individuals, but also from project teams and organizations;
- Have become a core competency of organizations, especially those that are project-based;
- Expand your study focus from a single project to the way organizations are using projects to achieve their goals;
- Have established a relationship with the execution of the organization's strategies;
- Consequently, it has become a competitive advantage for organizations.

It is worth remembering the concepts of project, program and portfolio constructs. A project is a temporary effort made to create a unique product or service.

A program is a group of projects, subprograms and related activities that are managed in a coordinated way to obtain benefits that could not be achieved if managed individually. For a program, centralized management and control allow both the operationalization of each one and the maintenance of the vision together of its objectives.

Finally, a portfolio is a set of projects, programs, sub-projects and operations, which are managed as a group to achieve the organization's strategic objectives. This alignment between the portfolio and the organization's strategic objectives is its main characteristic. In other words, the relationship between project management and an organization's strategy occurs through portfolio management.

The current project management approach comprises the basics of lean management, total quality management, re-engineering business process and the organization learning, among others.

In this context, if organizations need to manage their projects in an organized and systematized way, it is a fact that their processes need to evolve and mature. So, nothing is fairer than to realize this quest for excellence, through actions and planning, for the improvement of processes. That is, nothing more fair than seeking maturity in project management.

Therefore, the purpose of this chapter is to talk about what is maturity, what are the maturity models, and to present four of the major models currently in use.

## MATURITY IN PROJECT MANAGEMENT

There are several definitions of maturity in project management. Here are some of them.

Almeida (2005) makes a comparison between the characteristics of mature and immature organizations, as shown in Table 1.

According to Andersen and Jessen (2003), maturity is measured along three dimensions. They are:

- Knowledge, the ability to perform different tasks;
- Attitude, the will to achieve them;
- Action, the execution of it properly.

Thus maturity within an organization is explained as the sum of knowledge, attitudes and actions.

According to Costa and Ramos (2013):

*[...] project maturity can be understood as the use of appropriate practices by an organization, according to its area of performance, size, complexity and resources [...] the ability of an organization to coordinate these efforts - delivering projects at the same time as it directs its strategy - defines its maturity in managing organizational projects.*

On the other hand, Gareis (as cited in Costa & Ramos, 2013) associates maturity with the organization's competence in carrying out processes that it claims are critical to a project-oriented organization.

*Table 1. Mature and immature organizations*

| Mature Organizations | Immature Organizations |
|---|---|
| Well-defined roles and responsibilities | Improvised process |
| Existing historic base | There is no historical basis |
| It's possible to judge the quality of the process | There is no objective way to judge the quality of the process |
| Quality of products and processes is monitored | Quality and functionality of the product compromised |
| The process can be updated | There is no rigor in the process to be followed |
| There is communication between the manager and his team | Immediate crisis resolution |

Source: Almeida (2005).

Kerzner (2006) states that “the concept of maturity is defined as the development of systems and processes are inherently repetitive and guarantee a high probability that each one of them is a success”.

Thus, the word “mature” here means an environment where predictability is high and risk (the unknown) is low. Thus, analyzing the maturity of the process is assessing the balance between predictability and risk. The risk is reduced with increased predictability.

For PMI (2013a), “project management maturity [...] can be defined as the degree to which an organization practices managing organizational projects ... this is reflected in the combination of best practices achieved in project, program and portfolio domains “.

For Prado (2015), the concept of maturity in project management is linked to how skilled an organization is in managing its projects. For him, maturity produces significant results for organizations, such as:

- The reduction in the cost and term of development of new products;
- The increase in the life of new products;
- The increase in sales and revenue;
- The increase in the number of customers and their satisfaction;
- The increase of the chance of success in projects.

The above definitions open the way to various discussions about what maturity is. For the purpose of this chapter we consider a mature project management organization that:

- Has processes and procedures for project management established and used systematically;
- Has a cycle of continuous improvements acting on these processes and procedures;
- develops projects with high predictability and low risk;
- Uses project management as a competitive advantage to reach its strategies, mainly through the management of portfolios.

Once conceptually established what maturity is, comes the next question: How to evaluate or measure the maturity level of an organization? This is one of the roles of maturity models. Once the maturity status of an organization is known, improvement actions can be planned. This is another role of maturity models.

Thus, for Costa and Ramos (2013) the analysis of maturity is not only to understand the level in which the organization is. This analysis serves as a pillar for the establishment of objectives, actions and cultural change. In organizations, a common problem is to continue to focus on isolated projects and ignore the portfolio or the full spectrum of projects within the enterprise.

According to Pennpacker and Grant (as cited in Costa, Campos Filho & Lukosevicius, 2007), "maturity models provide a framework for evaluation that allows an organization to compare its project deliverables with best practices or against its competitors, defining a structured route for improvement."

The act of comparing with competitors clearly implies benchmarking actions, which are present in almost all models.

They also state that the application of maturity models in project management can take place at the sector (or department) level of an organization. For example, an organization's computer and engineering departments may have different maturity levels.

For the PMI (2013a):

*[...] maturity model is a conceptual framework, with constituent parts, that defines maturity in the area of interest - in this case, organizational project management. In some cases [...] maturity model may also describe a process by which an organization can develop or achieve something desirable, such as a set of capabilities or practices. This process can result in a more evolved organizational state; in other words, a more mature organization.*

Thus, a maturity model is a reference used to evaluate the ability of processes to achieve organizational goals; locate opportunities for productivity and quality improvement and cost reduction; plan and monitor the actions of continuous improvement of the business processes.

Several authors propose that the process of achieving maturity should be gradual; so that the planned actions can have their results evaluated in terms of their effectiveness and be absorbed by the organization. Thus, many of them structure the maturity of a company as a sequence of levels, which took the form of a ladder in many models.

Many maturity models assume that there are five levels. This has occurred since the first models appeared. Although they are equal in quantity, the meaning and content of each level changes from model to model.

Here are six examples of the models structured in five levels:

- Berkeley Project Management Process Maturity Model (PM)2, developed by Young-Honn KWAK and William C. IBBS;
- Capability Maturity Model (CMM) developed by the Software Engineering Institute (SEI) at Carnegie-Mellon University (CMU);
- Center for Business Practices, developed by PM Solutions;
- Maturity Model in Project Management (MMGP) developed by Darci PRADO;
- Project Framework, developed by ESI International;
- Project Management Maturity Model (PMMM) developed by Harold KERZNER.

However there are exceptions, such as:

- Model of Gareis and Hueman;
- Organizational Project Management Maturity Model (OPM3) developed by PMI.

The Maturity Model of Gareis and Hueman (2000) does not use ladder structure. The authors propose a spider web structure; because they understand that it is more flexible.

In this chapter our focus will be on the four maturity models that follow:

- Capability Maturity Model (CMM);
- Maturity Model in Project Management (MMGP);
- Organizational Project Management Maturity Model (OPM3);
- Project Management Maturity Model (PMMM).

Other models, in addition to these four, are mentioned in this chapter for contextual purposes only. These mentions may also serve as a starting point for readers who wish to go deeper into the subject.

## HISTORY OF MATURITY MODELS

Before we talk about maturity models properly, let's review a brief history of their emergence.

This history shows how maturity models are a relatively recent issue and also shows the contemporaneity of the models.

- Second World War: According to Verzuh (as cited in Costa & Ramos, 2013), project management evolved from major projects during and after World War II when the challenges were great and the normal techniques of project management were not adequate . Thus, it was in the period of war and post-war that the techniques of project management were defined and grouped, becoming a discipline in itself.
- 1979: Philip B. Crosby presented the Quality Management Maturity Grid, a maturity model for the adoption of quality concepts in an organization. It was structured in five levels.
- 1986: William E. Deming presented the practices for the continuous improvement of the quality management processes in an organization.
- 1989: Watts S. Humphrey extrapolated the concept of maturity of quality management processes to the "organizational" maturity measure, establishing a five-level maturity scale used to classify the software development process.
- 1990s: The recession in the middle of the decade led to many casualties in the North American organizations. Since then, the country's companies have begun to understand the importance of total quality management and the reduction of the development cycle of their products. Thus, companies have come to consider project management as a solution to some of their problems.
- August 1991: Launch of the first version (1.0) of the CMM model. The purpose of this model was specifically to gauge and promote the ability to manage software processes. It was developed by the Software Engineering Institute (SEI) of Carnegie Mellon University (CMU) in Pittsburgh, USA.
- 1993: Release of version 1.1 of the CMM model.
- 1996: CMM reviews interrupted due to CMM-I development prioritization.
- 1997: A. Fincher and G. Levin presented a maturity model based on CMM. He was one of the first to be adapted to project management.
- May 1998: Start of the PMI program (Project Management Institute) to develop a standard to help organizations measure and improve their project management capabilities and realize their organizational strategy. It was the beginning of OPM3 development.
- 1999: Start of the MMGP development by Darci Prado, based on his experiences in Brazilian companies.

- 2000: Release of the first version (1.02) of CMM-I (Capability Maturity Model Integrated), the successor to CMM, by SEI (Software Engineering Institute).
- 2001: Launch of the PMMM (Project Management Maturity Model) by Harold KERZNER.
- 2002: Young-Honn Kwak and William C. Ibbs presented the Berkeley Project Management Process Maturity Model (PM)2.
- December 2002: Launch of the first part of the MMGP by Darci PRADO. This first part proposes the sectorial evaluation, that is, the evaluation in the departments of companies and organizations.
- 2003: Launch of the organizational maturity model OPM3 by PMI.
- 2003: David HILLSON presented the maturity model ProMMM (Project Management Maturity Model). The structure of the ProMMM is divided into four levels. He was based on concepts and disseminated by CMM, EFQM (European Foundation for Quality Management) and a model that gauges organizational risk management capabilities (Risk Maturity Model).
- 2004: Launch of the second part of the MMGP (Maturity Model in Project Management) by Darci PRADO, which proposes the corporate evaluation, that is, of the organization as a whole. It extends its analysis to portfolio management and to sectors that are strongly linked to strategic goals being formally organized or not.
- 2008: Release of the second version of OPM3 by PMI, which was recognized by the American National Standards Institute (ANSI) as a US national standard.

## CAPABILITY MATURITY MODEL (CMM)

The CMM (Capability Maturity Model) was developed by the Carnegie Mellon University (CMU) Software Engineering Institute (SEI) in Pittsburgh, USA. Its first version (1.0) was released in August of 1991.

It was developed for specific application in the software development process. Thus, it is also named SW CMM (Software Capability Maturity Model).

It is based on process management concepts, quality and the experience of the software community of the North American industry.

This was backed by the US Department of Defense, which needed a formal model that would enable it to select its suppliers appropriately, as a major consumer of software. Thus, it originally had as main objective to classify the companies according to their abilities to design the software that would be used in the system of defense.

Although it was designed specifically to evaluate the maturity of software developers, it was the first maturity model (for project management) created.

The basic CMM philosophy can be described as a spiral of planning, implementation, and evaluation, with evaluation leading to improvements. After the improvements, cycle restarts. Thus, within this model we will find a PDCA cycle (plan, do, check and act). The presence of a PDCA cycle happens in practically all maturity models.

The CMM ended up giving rise to a family of maturity models, one for each activity of the software industry, such as: software acquisition, systems engineering and systems security engineering, among others.

In 2002 the SEI launched the CMM successor, the CMM-I (Capability Maturity Model Integrated). As the name suggests, it has integrated the various versions of the CMM.

## Maturity Levels

One of the purposes of the model is to describe the stages of maturity through which organizations pass as they evolve. This evolution is structured in five levels of maturity, each one characterized by a set of key areas whose structuring is considered necessary for the design and development of software.

The five levels and their main characteristics are described in Table 2.

Figure 1 graphically shows the structure of the model.

## Other Elements of the Model Structure: Key Process Areas

Each level of the CMM model is associated with a number of key process areas (KPA). The exception to this rule is the first level (Initial). Due to the disorganized and informal nature of your Level 1 processes, it has no key area.

Having mastery over the key areas is a prerequisite for reaching a certain maturity level and also a prerequisite for the next level. That is, if an organization is at Level 4 (Managed) it has mastery over the key areas of this level (4) and previous (2 and 3) as well.

Each key area is associated with key practices, which are activities that contribute to achieving the objectives of each area. The CMM model has

*Table 2. CMM maturity levels*

| Level | Name | Description |
|---|---|---|
| 1 | Initial | The process is disorganized and informal. Projects usually exceed the original deadlines and costs. Success depends on individual efforts and heroism. |
| 2 | Repetitive | Basic project management processes (functionality, schedule, and costs) are established. There is the discipline necessary to achieve success in projects similar to previous ones. |
| 3 | Defined | The standard of processes for software development exists. Management and engineering activities are documented, standardized, and integrated. All projects use these processes. |
| 4 | Managed | Quantitative control is established. Detailed measurements of process and product quality are carried out. That is, both the process and the product are understood and controlled quantitatively. |
| 5 | Optimized | The maturity model is institutionalized. Continuous improvement occurs through feedback of the process and the inclusion of innovative ideas and technologies. |

Source: Author

*Figure 1. CMM maturity levels*
*Source: Silva and Oliveira (2009).*

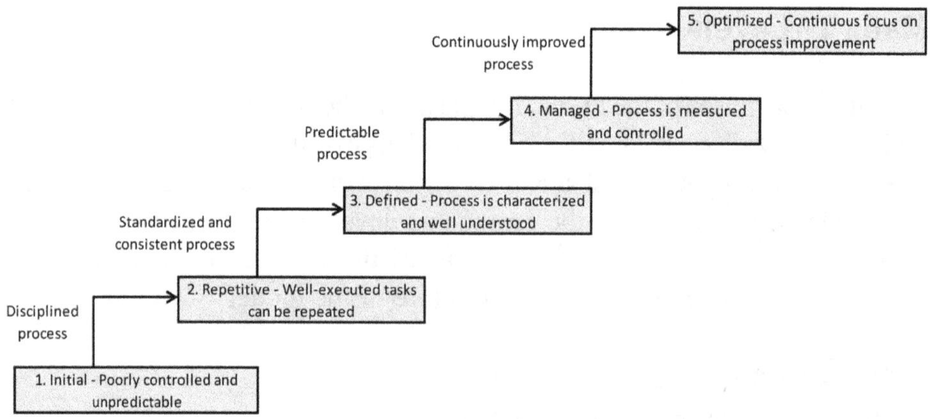


a separate document called "Base Practices for CMM", which lists all key practices for each of the key process areas.

For Level 2, key areas are focused on software design issues related to the establishment of basic project management controls. Are they:

- Monitoring and supervision of projects;
- Software quality assurance;
- Requirements management;
- Subcontracting management.

- Configuration management;
- Project planning.

At Level 3, they are focused on both project and organization issues as the organization establishes an infrastructure that effectively institutionalizes software engineering and project management processes.

- Coordination between groups;
- Definition of the organization process;
- Software product engineering;
- Focus on the organization process;
- Integrated software management;
- Training program;
- Reviews.

For Level 4, the focus is on the quantitative establishment of both the software process and the products under construction.

- Software quality management;
- Quantitative process management.

At Level 5, they cover issues that both the organization and the projects must consider to implement software improvements that are continuous and measurable.

- Process change management;
- Management of technological changes;
- Prevention of defects.

## Model Application

Through an extensive questionnaire, follow-up interviews, and evidence gathering, software organizations could be "ranked" at one of five maturity levels, based primarily on the rigor of their development processes.

## MATURITY MODEL IN PROJECT MANAGEMENT (MMGP)

The MMGP model was developed by Darci Prado between 1999 and 2002, based on his experiences in Brazilian companies, involving hundreds of projects. It addresses maturity in project management from two perspectives: the sectorial and the corporate.

The MMGP sectorial was launched in December 2002. It aims to classify the organization's ability to manage its projects in five levels and traces its profile in seven dimensions in a departmental way. That is, sector model applies to isolated sectors (or departments) of an organization, such as information technology and engineering.

The corporate MMGP was launched in 2004. It aims to evaluate the organization as a whole and extends its analysis to portfolio management and to sectors that are strongly linked to strategic goals being formally organized or not.

Like other maturity models, the MMGP uses the constructs of project, program and portfolio, which establishes a relationship between itself and the PMBOK (Project Management Book of Knowledge) of the PMI (Project Management Institute).

Our focus here is the sectorial MMGP. One of the basic characteristics of this model is to be results-oriented, so high-performance departments are expected to achieve high maturity values.

The MMGP seeks to be simple to use, to provide reliable and robust results. According to Prado (2005) "the complex models would eventually alienate their potential users, making the subject matter mature, but little practiced". Thus, Prado (2005) enumerates the following assumptions used in the development of the MMGP:

- Being small (the questionnaire contains only 40 questions);
- Be simple to use;
- Be trustworthy;
- Provide consistent results (robust);
- Have universality (be able to be used by different categories of projects);
- Be able to measure aspects that are really linked to success in project management;
- Conquering adepts for maturity;
- Can be used in establishing a maturity development plan.

This model differs from other models because it does not present the need to invest in specific software to compile the data.

Further information on MMGP is available at www.maturityresearch.com, including up-to-date data from surveys conducted in Brazil (since 2005) and in other countries. Among these data there is a database for benchmarking.

## Maturity Levels

Similar to other models, MMGP has five levels of maturity. When analyzing the levels and their characteristics we observed the presence of the processes prioritized by Prado in the development of the model, which are:

- Existence and use of good project management practices;
- Elimination of the causes of the anomalies;
- Continuous Improvement;
- Permanent technological and process innovation;
- Sustainability.

The five levels and their main characteristics are described in Table 3.

Remember that the industry model applies to isolated sectors (or departments) of an organization, such as information technology and

*Table 3. Maturity levels of the MMGP model*

| Level | Name | Description |
|---|---|---|
| 1 | Initial, embryonnaire or *ad hoc* | The organization has not yet made any effort to implement a project management model. Projects are executed on the basis of intuition, goodwill and individual effort. Success is the fruit of individual effort or chance. |
| 2 | Known | The organization has awakened to the subject of project management. An effort was made to create a language for the subject of project management. This effort was made from different types of training. |
| 3 | Standardized | The organization has deployed and uses a standardized project management model based on competencies, processes and tools. A management platform was deployed. |
| 4 | Managed | The organization has consolidated the efficient and effective practice of the model implemented in Level 3 (organizational structure, methodology, computerization and strategic alignment). The organization uses really competent professionals. It also consolidates a strong alignment of projects with company strategies. |
| 5 | Optimized | The organization has attained the wisdom of project management. All initiatives implemented at previous levels have reached a level of excellence. The organization practices continuous improvement and technological and process innovation. |

Source: Author based on Prado (2015)

engineering. Thus, an organization may have sectors with different levels of maturity. Eventually, an organization may be in Level 3 in the IT sector and in Level 2 in the engineering sector.

## Other Elements of the Structure of the Model: Dimensions

In addition to the five maturity levels, the MMGP model is based on seven dimensions. They are:

- Competence in project management, program and portfolio: people involved with project management must be proficient (knowledge and experience) in project management.
- **Behavioral Competence:** The people involved should be competent in behavioral aspects. This should encompass knowledge of human relationships, in order to motivate team members and minimize possible conflicts.
- **Technical and Contextual Competence:** The people involved must be competent in technical aspects related to the product being created and with aspects related to the organization.
- **Methodology:** A methodology should be structured and used for project management. It should encompass a number of steps to be followed to ensure the correct application of methods, techniques and tools.
- **Informatization:** The methodology must be computerized through an easy-to-use system. It must allow access to the various projects, their performance indicators and must support decision-making at the right time.
- **Organizational Structure:** An organizational structure should be established encompassing project managers, PMO (Project Management Office), sponsors and committees. It should regulate the relationship of authority and power between project managers and the various areas of the organization in order to reduce conflicts and increase results.
- **Strategic Alignment:** Current and future projects are in total alignment with the organization's strategies. Portfolio management is executed with the necessary quality and agility.

The "structural" arrangement of these seven dimensions, that is, how they relate is shown in Figure 2.

*Figure 2. The seven dimensions of maturity in the MMGP*
*Source: Prado (2015)*

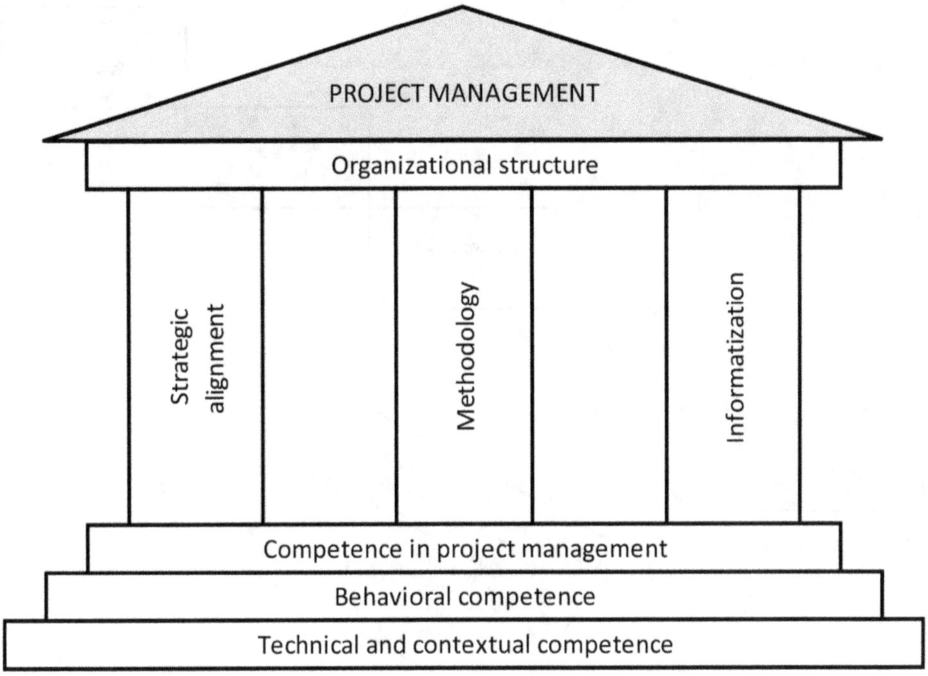


The relationship between maturity levels and dimensions, i.e., the complete structure of the MMGP model is shown in Figure 3.

Each level contains the seven dimensions of maturity, in different degrees of maturity or optimization, depending on the characteristics of each level. The exception is Level 1 in which, due to the disorganized and informal nature of its process, none of the dimensions are observed. In Table 4 we can see how the seven dimensions are characterized at each maturity level.

## Model Application

The Sectorial MMGP Model should be applied separately in each sector (or department) of an organization. The components of the model for evaluation and improvement are three:

- A questionnaire of 40 questions;
- Guidelines for making a diagnosis (simplified or detailed);
- Guidelines for creating a maturity development plan.

*Figure 3. Model MMGP Structure*
*Source: Prado (2015)*

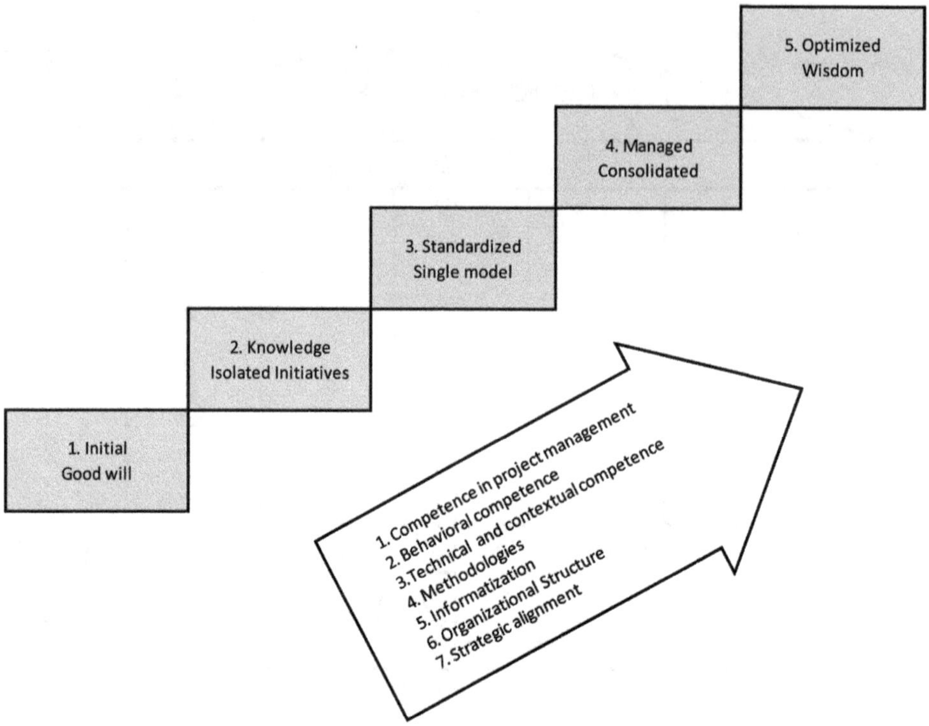


*Table 4. Characteristics of maturity dimensions according to the levels*

| Level | Maturity Dimensions | | | | | | |
|---|---|---|---|---|---|---|---|
| | **Competence in Project Management** | **Technical and Contextual Competence** | **Behavioral Competence** | **Methodology** | **Informatization** | **Organizational structure** | **Aligned** |
| 5 | Optimized | Optimized | Mature | Optimized | Optimized | Optimized | Significant advance |
| 4 | Much advanced | Strong advance | Strong advance | Improved, stabilized and in use | Improved, stabilized and in use | Improved, stabilized and in use | Non-aligned |
| 3 | Significant advance | Some advance | More advance | Standardized and in use | Standardized and in use | Standardized and in use | Non-aligned |
| 2 | Basic | Basic | Some advance | Discuss about the subject | Software for time | Assign a responsible | Aligned |
| 1 | Inexistent | Basic | Good will | inexistent | Disperse | Inexistent | Significant advance |

Source: Prado (2005)

## Questionnaire

The questionnaire, divided into four stages, is characterized by simplicity (questionnaire of 40 multiple choice questions) and universality (applicable to any type of organization). The questionnaire should be answered by the person in charge of the sector under evaluation.

The questionnaire presents five alternative responses (from "A" to "E") for each question. From the answers, it is possible to evaluate the maturity of this sector resulting in three different information:

- Final evaluation of maturity, which represents the level of maturity of the sector evaluated;
- Adherence to levels;
- Adherence to dimensions.

The Final Maturity Assessment is obtained through:

- From the assignment of values to the answers, according to the following scale: A = 10, B = 7, C = 4, D = 2 and E = 0;
- From the sum of the points obtained;
- From the application of the following formula: (100 + "total points") / 100.

The result will be between 1 and 5.

The percentages of adherence to levels and dimensions use the same score, but with the sum done in each of the four stages of the questionnaire. These steps are permeated by the requirements of each maturity level. Then, the result being:

- Null or weak adhesion: up to 20%;
- Regular adhesion: from 20% to 60%;
- Good adhesion: from 60% up to 90%;
- Full adhesion: over 90%.

## Guidelines for Diagnosis

The Diagnosis is to deepen the understanding of the current situation and the causes of the weaknesses found in the sector. It is a prerequisite for the maturity development plan, the Growth Plan.

This phase includes benchmarking. The research carried out by Darci Prado and Russel Archibal can be used, which presents information on the average level of maturity of Brazilian organizations from the year 2005. This information is available on the website www.maturityresearch.com.

Another aspect to be evaluated in the diagnosis is the time of use of the new practices. The model requires that after implementation of some aspect of management, it is practiced by all involved for a minimum period of time to be accepted that the aspect is consolidated. The deadlines are as follows:

- Levels 2 and 3: Minimum of 12 months;
- Levels 4 and 5: Minimum of 24 months.

The model also requires that a significant number of projects must undergo all (new) processes of the methodology.

### Growth Plan Guidelines

The Growth Plan Guidelines were created to enable maturity evolution as a project, with all the peculiarities that a project and its management should have. It is certain that the Growth Plan will involve other sectors of the organization besides the sector evaluated.

## ORGANIZATIONAL PROJECT MANAGEMENT MATURITY MODEL (OPM3)

Released in 2003 by the PMI (Project Management Institute), the OPM3 (Organizational Project Management Maturity Model) expands from the project domain, the main object of the Project Management Book of Knowledge (PMBOK), to also encompass program and portfolio domains.

This model was developed through the participation and consensus of a diverse group of project management professionals spanning several organizations in thirty-five countries.

Thus, OMP3 aims to be applicable in cultures around the world and in virtually any industry, from engineering and construction to information technology, financial services, government agencies and manufacturers, to name a few.

According to the PMI (2013a), the use of OPM3 brings the following benefits to organizations:

- Ability to promote the right projects, in the right way, strategically aligned with the organization;
- Flexibility to be used by different types of organizations, through different areas of activity, business, size, geographical, etc.;
- Awareness and clarification of the issue of organizational maturity for all senior management;
- Establish an association between the success of the organization and the effective and efficient management of projects.

The differential of OPM3 is that it does not use the classification structure in levels. Its structure is designed in four dimensions:

- **Best Practices and Capabilities:** The model covers a set of nearly 600 best practices. The model also defines what capabilities need to be developed so that practices are recognized as implemented.
- **PPP Domain:** Relates best practices (and their capabilities) to the specific domain for projects, programs or portfolios. This is the PPP domain.
- **SMCI Internships:** Relates best practices to the organization's maturity stage. Stages of maturity are: standard, measured, controlled and improvement. These are the SMCI stages.
- **IPECC Process Group:** Lists best practices with the project management process group, according to PMBOK: initialization, planning, execution, controlling and closing. These are the IPECC process groups.

For this model:

*[...] maturity is commensurate with the ability to perform each of the processes well and implies the existence of relevant best practices. Maturity also includes the concept of establishing project-level standards, process measures, process controls, and continuous process improvement. The existence or attainment of maturity in any process depends on the existence of various capabilities. The ability to control an input or process depends on the ability to standardize that input or process and the ability to measure the stability of that input or process (PMI, 2013a).*

Its use is quite complex and depends on a specific software, ProductSuite OPM3®, which is a set of tools for diagnosis and improvement. PMI-specific certification is required for service providers and consultants to use it.

## Stages of Maturity and Domains

OPM3 was designed intentionally without a general system of "levels" of maturity. Establishing specific levels of maturity can be relatively simple if the maturity progression is one-dimensional. That is, OPM3 does not use the tlevels structure. Its structure is designed in four dimensions:

- Best practices and capabilities;
- PPP domains;
- SMCI internships;
- IPECC process group.

In a superficial analysis we can associate the maturity stages (SMCI) with the maturity levels of other models, but this comparison does not hold when we include in the analysis the other dimensions of OPM3.

The PPP domain and the SMCI stages are related as shown in Figure 4. Remember that this figure does not represent the entire structure of the model because it covers only two of the four dimensions.

According to the PMI (2013a):

*Multiple perspectives to assess maturity allow flexibility in applying the model to the unique needs of an organization. This approach also produces a more robust set of information than is possible with a simpler, more linear level system, giving the organization greater detail in support of decisions and improvement plans.*

## Other Elements of the Model Structure: Best Practices and Abilities

The best practices and capabilities in the OPM3 are intentionally of general nature in order to cover the project management and variety of products.

A best practice is the best way to achieve an specific objective. For the projects management, this also includes the ability to deliver projects in a predictable way, consistent and aligned with the organization's strategies.

*Figure 4. Relationship between the maturity stages and the PPP domain in the OPM3*
*Source: PMI (2013a)*

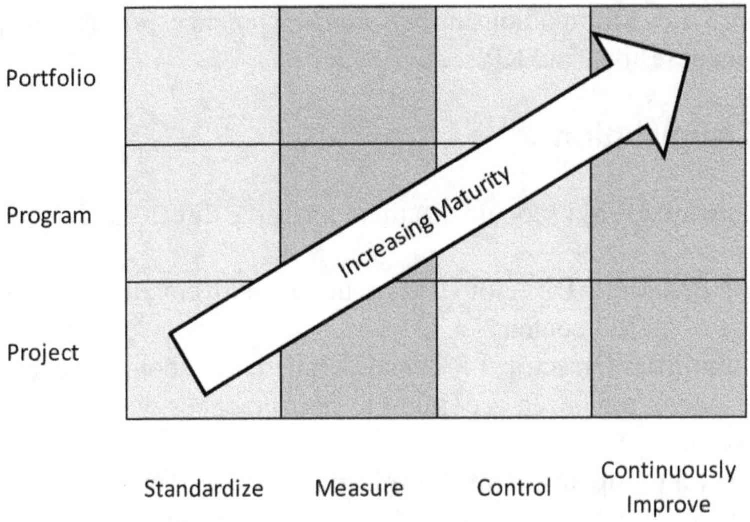


Besides that, the best practices are dynamic because evolve with time as new and better approaches are developed.

The best practices in OPM3 cover an ample category spectrum, such as:

- Allocate resources for the projects;
- Control and improve projects continuously;
- Develop team's competences of project management;
- Develop appropriate governance structures;
- Desenvolver o compromisso com a gestão de projetos;
- Improve team work;
- Standardize and integrate processes;
- Prioritize projects and align them with the organizational strategy;
- Utilize success criteria to continue or to close projects;
- Utilize performance metrics.

Each best practice consists of two or more capabilities. A capability is a specific competency that must exist within an organization for it to execute project management processes properly.

The existence of a capacity is demonstrated by the existence of one or more corresponding results. In OPM3, one capability can have multiple results. Thus, a key performance indicator (KPI) is a criterion by which an organization can determine whether the result associated with a capability has been achieved, even partially.

Figure 5 shows the relationship between best practices, capabilities, project management results, and KPIs.

## Model Application

In the standard OPM3 (document) there are three directories:

- **Best Practices Directory:** Lists nearly 600 practices that form the basis of OPM3 content.
- **Capabilities Directory:** Provides detailed data about all capabilities in the model, organized according to best practices with which they are associated.
- **Directory Improvement Planning:** Shows the dependencies between capabilities, which are essential for evaluation and improvement within the model.

The model provides the organizations with tools for them to evaluate and the means to decide whether or not to pursue through an improvement plan. For those who choose to make improvements, OPM3 provides steps to determine an appropriate course of action as shown in Table 5.

To support the evaluation process PMI created the ProductSuite OPM3®, which is a set of software tools that allows for maturity evaluations according to the OPM3 model.

Only PMI Certified Consultants can use the ProductSuite OPM3®.

To finalize the discussions on this model, it is worth mentioning that Andersen and Jessen (2003) developed a field study with 59 project managers, using the OPM3 model and its domains (project, programs and portfolio)

*Figure 5. Relationship between best practices, capabilities, project management results and performance indicators in OPM3*
*Source: PMI (2013a).*

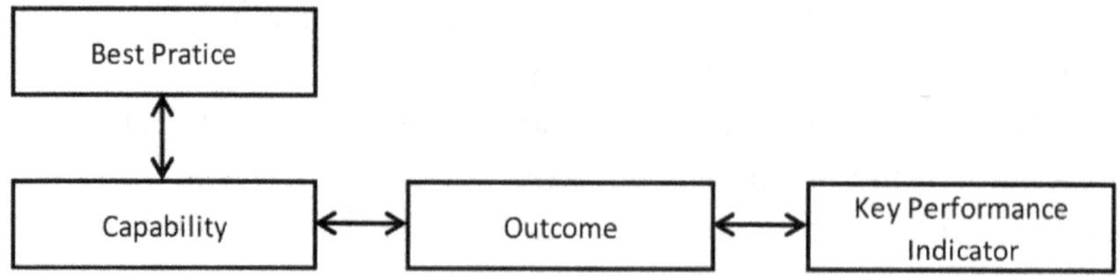

*Table 5. Stages of evaluation and improvements in OPM3*

| Stage | Name of the Stage | Description |
|---|---|---|
| 1 | Prepare the evaluation | It covers the understanding of the content of the model as completely as possible. The organization must become familiar with project management and the operation of OPM3. |
| 2 | Carry out the evaluation | The maturity stage evaluation is divided into two stages. The first is a capacity survey based on a questionnaire of 151 questions that is used in the production of 4 graphs that indicate in general the maturity of the organization against the standard. In the second stage, the focus of the organization is the study of capabilities. |
| 3 | Elaborate improvement plan | It consists of the elaboration of an improvement plan based on the best practices and capacities identified as undeveloped within the domain and stage of maturity of interest. |
| 4 | Implement improvements | Implementation of the improvement plan that is, implementing organizational development activities necessary to achieve the necessary capabilities to increase maturity in project management. |
| 5 | Repeat process | After have implemented, even partially the improvement plan, the organization returns to the first step (recommended) or the third step to address other best practices identified in an earlier evaluation. |

Source: Author based on the PMI (2013a)

and concluded that there are indications that maturity at the base (individual project management) is higher than at the top (portfolio management).

## PROJECT MANAGEMENT MATURITY MODEL (PMMM)

The Project Management Maturity Model (PMMM) was presented by Harold Kerzner in 2001.

Like other maturity models, PMMM is structured in five levels. It also uses benchmarking to measure the progress of an organization throughout the model.

In addition, PMMM uses PMOBOK's nine areas of knowledge and project, program and portfolio constructs, which establishes a relationship between PMBOK (Project Management Book of Knowledge) of PMI (project Management Institute).

### Maturity Levels

The PMMM proposes that for a company to achieve excellence in project management, it is necessary to advance through the five levels shown in Table 6.

*Table 6. PMMM maturity levels*

| Level | Name | Description |
|---|---|---|
| | Common language | The organization recognizes the importance of project management and the need to have a knowledge base to at least be able to establish terminology. |
| | Common processes | The organization recognizes that common processes need to be defined and developed so that the success of one project can be replicated to everyone else in it. |
| | Singular methodology | The organization recognizes the synergy generated by the combination of all corporate methodologies in a single one, whose central axis is project management. |
| | Benchmarking | The organization recognizes that process improvement is necessary to maintain a competitive advantage. Benchmarking should be done on an ongoing basis. |
| | Continuous improvement | The organization evaluates the information obtained from benchmarking and decides whether they can improve their methodology in project management. |

Source: Author.

It is interesting to note that the CMM (Capability Maturity Model) and MMGP (Project Management Maturity Model) models consider Level 1 to be disorganized and informal.

In a different way, the PMMM considers a minimum structuring in Level 1, since it admits the existence of a common language, which would put it above Level 1 of the other two models.

Thus, can it be suggested that Level 1 of the CMM and MMGP would correspond to a hypothetical Level 0 in the PMMM?

Figure 6 shows the structure of the PMMM model, highlighting the continuous improvement process that occurs from Level 5, but covers Levels 3 and 4 as well.

Overlap between levels may occur. Despite this, the order in which the levels are completed does not change. For example, even if Level 1 and Level 2 can overlap, Level 1 must still be completed before Level 2.

The process of reaching, or completing, each level of PMMM has risks. They can be labeled as low, medium and high. The level of risk is often associated with the impact it can have on corporate culture. As shown below:

- **Low Risk:** Practically no impact on corporate culture, or corporate culture is dynamic and readily accepted changes.
- **Medium Risk:** The organization recognizes that change is necessary but may not know the impact of change.

*Figure 6. PMMM structure*
*Source: Kerzner (2001)*

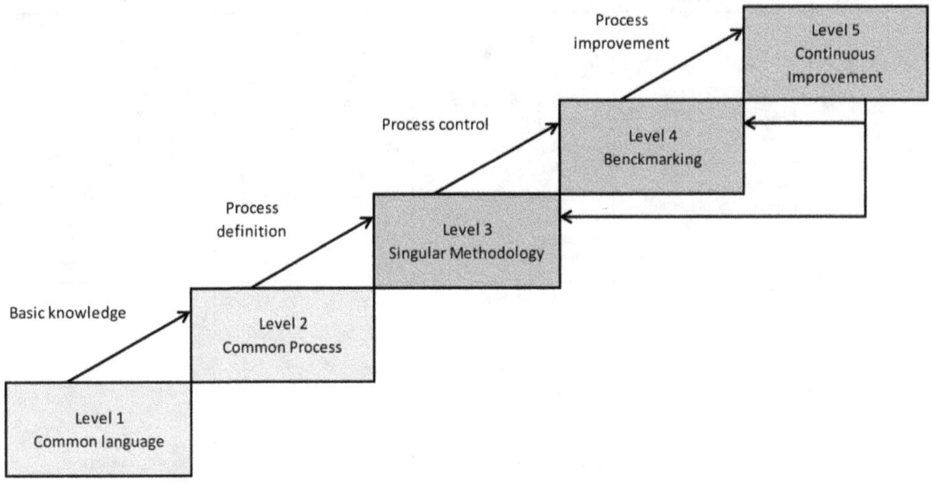


- **High Risk:** Occurs when the organization recognizes that changes resulting from the implementation of project management will cause a change in corporate culture.

In Table 7 we related the blockages, progress criteria (for reaching or completing the level) and risks associated with the first four levels of the model according to Kerzner (2001).

## Other Elements of the Structure of the Model: Phases of Level 2

According to Kerzner (2006), it is in Level 2 that the maturity of the organization has its true beginning. It begins the search for the common processes that will later be the basis of the unique methodology of the organization.

According to the model, the search for these common processes can be structured as a five-stage cycle, as shown in Table 8.

## Model Application

The model establishes a specific evaluation process for each of the maturity levels.

*Table 7. Blockages, advancement criteria and associated risks according to the levels*

| Level | (1) Common Language | (2) Common Processes | (3) Singular Methodology | (4) Benchmarking |
|---|---|---|---|---|
| Blockroads | • Resistance to change.<br>• Leaving well enough alone.<br>• Not invented here.<br>• It does not apply to us.<br>• We don't need it. | • Resistance to a new methodology.<br>• What we already have works well.<br>• Believing that a methodology needs rigid policies and procedures<br>• Resistance to "horizontal" accounting. | • Don't fix it if it isn't broken.<br>• Resistance to a singular integrated methodology (i.e., repeatable process).<br>• Resistance to shared accountability.<br>• Fragmented corporate culture.<br>• Overemphasis on documentation. | • "Not invented here" syndrome.<br>• Does not apply to us.<br>• Wrong industry to benchmark against.<br>• Fearful of what results will be found.<br>• Resistance to change. |
| Advancement criteria | • Recognize available project management tools.<br>• Develop an understanding of the principles of project management: the project management body of knowledge (PMBOK). | • Develop a culture that supports both the behavioral and quantitative sides of project management.<br>• Recognize both the driving forces/need for project management and the benefits that can be achieved in both the short and the long term.<br>• Develop a project management process/ methodology such that the desired benefits can be achieved on a repetitive basis.<br>• Develop an ongoing, all-employee project management curriculum such that the project management benefits can be sustained and improved upon for the long term. | • Integrate all related processes into a single methodology with demonstrated successful execution.<br>• Encourage the corporate-wide acceptance of a culture that supports informal project management and multiple-boss reporting.<br>• Develop support for shared accountability. | • Create an organization dedicated to benchmarking.<br>• Develop a project management benchmarking process.<br>• Decide what to benchmark and against whom to benchmark.<br>• Recognize the benefits of benchmarking. |
| Risks | MEDIUM<br>• Fear of organizational restructuring.<br>• Fear of changes in roles and responsibilities.<br>Fear of changes in priorities. | MEDIUM<br>• Type of company (project-driven versus non–project-driven).<br>• Visibility of executive support.<br>• Strength of the corporate culture.<br>• Resistance to change.<br>• Speed with which a good, workable methodology can be developed.<br>• Existence of an executive-level champion to drive the development of the project management methodology.<br>Speed with which the project management benefits can be realized. | HIGH<br>• The speed at which the culture will change.<br>• The acceptance of informal project management.<br>The acceptance of a singular methodology | LOW<br>The successful completion of Level 4 is accompanied by a low degree of difficulty. Since the organization has already accepted the idea of a singular methodology, it is a low risk to expect the employees to accept changes. They now know that change is inevitable |

Source: Author based on Kerzner (2001)

*Table 8. Phases of level 2 (common process)*

| Phase | Name | Description |
|---|---|---|
| 1 | Embryonic | It is when you see whether there is recognition (or not) of the importance of project management for the organization. |
| 2 | Executive Management Acceptance | This is when top management recognizes the importance of project management. This manifests itself through visibility in terms of support, understanding of the discipline of project management, establishment of sponsors at the executive level and favorable posture in changing the way of doing business. |
| 3 | Line Management Acceptance | This is when line management supports and is committed to project management. This is manifested in fulfilling the objectives and in releasing the necessary resources to the process. |
| 4 | Growth | It is when the methodology of project management is disseminated, promoting the recognition of its importance for the activities purposes and in the diverse functions of the organization. |
| 5 | Maturity | It is when a formal management control system is established that integrates costs and deadlines, as well as the development of an educational program with the purpose of increasing the skills in project management in the organization. |

Source: Author

For Level 1 (Common Language) an 80-question questionnaire covering all the main areas of knowledge of the PMBOK is applied.

In Level 2 (Common Processes) a 20-question questionnaire provides the status of the organization in the common process search cycle. Each question has seven possibilities of answer, each answer is then assigned a number ranging from -3 to +3, as follows:

- (-3) I strongly disagree;
- (-2) Disagree;
- (-1) Partially disagree;
- (0) No opinion;
- (1) I agree partially;
- (2) I agree;
- (3) I totally agree.

The value of the answers is placed in the table below, where the calculated totals will show in which phase the organization is.

A high score (usually equal to or greater than +6) indicates that those phases of evolution for the Singular Methodology have already been achieved. Phases with very low scores indicate that this was not achieved.

*Table 9. Assessment of Level 3 phases*

| Embryonic | | Executive Management Acceptance | | Line Management Acceptance | | Growth | | Maturity | |
|---|---|---|---|---|---|---|---|---|---|
| Question | Value | Question | Value | Question | Value | Question | Value | Question | Value |
| 1. | | 5. | | 7. | | 4. | | 2. | |
| 3. | | 10. | | 9. | | 6. | | 15. | |
| 14. | | 13. | | 12. | | 8. | | 16. | |
| 17. | | 20. | | 19. | | 11. | | 18. | |
| Total: | | Total: | | Total: | | Total: | | Total: | |
| Grand Total: | | | | | | | | | |

Source: Kerzner (as cited in Costa & Peinado, 2003).

Already for Level 3 (Singular Methodology) a questionnaire of 42 questions assess the six characteristics of the so-called "hexagon of excellence": integrated processes, culture, support and support of leadership, training and education, reduction of bureaucracy in project management.

In Level 4 (Benchmarking), it assesses the extent to which an organization makes use of the benchmarking process and benchmarking practices to improve its project management.

For Level 5 (Continuous Improvement) evaluates the processes and practices adopted by the organization to safeguard, consolidate, improve and disseminate lessons and accumulated learning with the execution of project management in the organization.

## CONCLUSION

In this chapter we talked about what is maturity in project management, what are maturity models, what are their roles and presented a brief history of the emergence of these models. We also analyzed in detail the following maturity models:

- Capability Maturity Model (CMM);
- Maturity Model in Project Management (MMGP);
- Organizational Project Management Maturity Model (OPM3);
- Project Management Maturity Model (PMMM).

In the analysis of these models, especially the last three, some points in common stood out, even if present in different intensities or degrees. Some of them are listed below:

- Structure in five levels.
- Use of benchmarking.
- Use of project, program and portfolio construct.
- Use of PMBOK process groups: initialization, planning, execution, controlling and closing.
- Explicit relationship between project management and the organization's strategy.
- Existence of PDCA cycles (plan, do, check and act) within the model, covering one or more levels.

On the five-level structure, it is observed that with the exception of OPM3, the other models analyzed work with five levels of maturity. This is probably influenced by the CMM (the first of the models), which in turn was influenced by the works of Humphrey.

Recall that in 1989 Watts S. Humphrey extrapolated the concept of maturity from quality management processes to the measure of "organizational" maturity, establishing a five-level maturity scale.

Still in relation to maturity levels, we have a difference of PMMM when compared to CMM and MMGP.

The CMM and MMGP models consider that Level 1 is disorganized and informal. In a different way, the PMMM considers a minimum structuring in Level 1, since it admits the existence of a common language.

Thus, can it be suggested that Level 1 of the CMM and MMGP would correspond to a hypothetical Level 0 in the PMMM?

On the projects, programs and portfolios construct, it is observed that with the exception of the CMM, the analyzed models work use this construct. They also use other PMBOK concepts and processes. Obviously this relationship is stronger in OPM3, which was also developed by PMI.

Regarding the relationship between project management and the organization's strategy, it is observed that the models try to establish this by explicitly mentioning the management of portfolios. This relationship varies from model to model, and seems to be more explicit in OPM3 and PMMM.

Each model has its characteristics or peculiarities. Just to name a few:

- CMM was developed especially for the software industry.
- MMGP proposes to be an easier application model, since the simple questionnaire that it uses and does not need specific software to apply it.
- OPM3 has the most complex structure (four dimensions and almost six hundred best practices).
- The PMMM proposes to have a strong relationship with the organizations' strategy.

Which model should a particular organization choose? We propose that the choice to be made by the model whose characteristics are more aligned with the strategies and culture of the organization, and why not say, in line with the peculiarities of it as well.

Obviously there are other models that were not detailed here, such as the model of Gareis and Hueman (2000) that reject the structure of levels and use a structure of spider web.

The purpose of this chapter was to present four of the major maturity models, to establish the context, relative to the models, for the subsequent discussions of this book. This goal was achieved.

In addition, this chapter can serve as a starting point for those who wish to delve into the subject of maturity models for project management, programs and portfolios.

## REFERENCES

Almeida, K. (2005). As Características do CMM e o desenvolvimento de software com qualidade [The Characteristics of CMM and the development of software with quality]. *Revista eletrônica de administração, 8*. Retrieved from <http://faef.revista.inf.br/imagens_arquivos/arquivos_destaque/2NZmUIDpT4zWfx8_2013-4-26-9-45-49.pdf>

Andersen, E. S., & Jessen, S. A. (2003). Project maturity in organizations. *International Journal of Project Management, 21*(6), 457–461. doi:10.1016/S0263-7863(02)00088-1

Bouer, R., & de Carvalho, M. M. (2005). Metodologia singular de gestão de projetos: Condição suficiente para a maturidade em gestão de projetos? [Singular methodology of project management: sufficient condition for maturity in project management?]. *Production, 15*(3), 347–361. doi:10.1590/S0103-65132005000300006

Costa, H. G., Campos Filho, L. A. N., & Lukosevicius, A. (2007). Maturidade em gerenciamento de projetos e desempenho dos projetos [Maturity in project management and project performance]. *Relatórios de Pesquisa em Engenharia de Produção, 7*(7).

Costa, S. R., & Ramos, A. F. B. (2013). Modelo de Maturidade em Gerenciamento de Projeto: Um Estudo de Caso Aplicado a Projetos de Petróleo e Energia [Maturity Model in Project Management: A Case Study Applied to Oil and Energy Projects]. *Sistemas & Gestão, 8*(3), 234–243. doi:10.7177g.2013.v8.n3.a2

Da Costa, J. H. C., & Peinado, J. (2003). *Análise da maturidade em gestão de projetos: aplicação do modelo PMMM em fornecedores da indústria automobilística* [Analysis of the maturity in project management: application of the PMMM model in suppliers of the automobile industry]. Academic Press.

Gareis, R., & Huemann, M. (2000). *Project management competences in the project-oriented organization. In The Gower handbook of project management* (pp. 709–721). Aldershot, UK: Gower.

Kerzner, H. (2001). *The project management maturity model.* New York: John Wiley & Sons.

Kerzner, H. (2006). *Gestão de Projetos: As Melhores Práticas* [Project Management: Best Practices]. São Paulo, Brazil: Bookman Publisher.

Marisco, T. (2009). *Avaliação da maturidade em gerenciamento de projetos: estudo multicaso de empresas globais no contexto brasileiro (Monografia de Especialização)* [Maturity evaluation in project management: a multi-stage study of global companies in the Brazilian context (Specialization Monograph)]. Belo Horizonte, Brazil: Ibmec Univesity.

Monteiro de Carvalho, M., Rabechini Jr, R., Schneck de Paula Pessôa, M., & Barbin Laurindo, F. J. (2005). Equivalência e completeza: análise de dois modelos de maturidade em gestão de projetos [Equivalence and completeness: analysis of two maturity models in project management]. *Revista de Administração-RAUSP, 40*(3).

Moura, A. A. (2013). *Gerenciamento de riscos associados a stakeholders em projetos de mineração. (Monografia de especialização)* [Management of risks related to stakeholders in mining projects. (Specialization Monograph)]. Belo Horizonte, Brazil: Pitágoras University.

Persse, J. R., & Persse, D. J. (2001). *Implementing the capability maturity model.* New York: John Wiley & Sons, Inc.

Prado, D. (2015). *Maturidade em gerenciamento de projetos* [Maturity in project management]. 3rd ed.). Nova Lima, Brasil: FALCONI Editora.

Project Management Institute. (2013). Organization Project Management Maturity Model (OPM3) (3rd ed.). Project Management Institute.

Project Management Institute. (2013). *Project Management Body of Knowledge (PMBOK)* (5th ed.). Project Management Institute.

Segismundo, A., & Carvalho, M. (2006). *Maturidade em gestão de projetos: análise comparativa em 3 unidades de negócio do setor automobilístico* [Maturity in project management: comparative analysis in 3 business units of the automotive sector]. XXVI ENEGEP.

Silva, L. P. C., & Oliveira, L. C. A. (2009). *Modelo CMM: Uma Visão Geral* [CMM Model: An Overview]. Curitiba, Brasil: Bate Byte.

Silva Neto, J. C. (2011). *Avaliação de maturidade no gerenciamento de projetos em uma empresa de mineração em minas gerais* [Evaluation of maturity in the project management in a mining company in Minas Gerais State] (Masters dissertation). Fundação Mineira de Educação e Cultura, Graduate Program in Administration, Belo Horizonte, Brazil.

Soler, A. M. (2002). Maturidade Organizacional e o Modelo de Avaliação PMI OPM3 [Organizational Maturity and the Evaluation Model. PMIOPM3]. Rio de Janeiro: Campus.

Zaguir, N. A., & Martins, M. R. (2007). Revisão Crítica do OPM3: Um estudo de redundâncias [Critical Review of OPM3: A study of redundancies]. *Revista Gestão Industrial, 3*(1).

# Section 3

# A Model to Measure Complex Projects

# Chapter 7
# Choice of Maturity Model to Complex Engineering Projects


## ABSTRACT

*After understanding maturity in project management, it is important to understand how to choose the ideal model for complex engineering projects. This choice will be assigned to the number of interfaces in the project. Survey of stakeholders and how they relate to the enterprise will directly result in the number of such interfaces. Once the maturity model is understood, it will be possible to manage the projects more accurately and increase the chances of success. This chapter will present how interfaces are identified and managed in an engineering and construction project.*


## INTRODUCTION

This paper discusses the reasons that justify the use of Interfaces Management through a multidisciplinary coordination in industrial projects. It presents a real application case and practical change management. In addition, it makes a recommendation of processes to carry out this management among the several disciplines that compose the industrial projects focusing on the engineering phase (design) of the same. Based on this information, it presents the benefits that the use of the interface management can bring to the industrial enterprises guaranteeing the choice of an ideal maturity model allowing a better management of complex projects.

DOI: 10.4018/978-1-5225-5864-4.ch007

Projects for the construction or renovation of industrial plants are usually large and expensive efforts for companies with a high degree of difficulty and complexity. In this way, Project Management techniques are used to increase their chances of success, through a chained and organized administration.

The large number of companies involved in such ventures makes it even more difficult to coordinate the activities that influence and relate, since it must be ensured that information is available as accurately as possible and in a timely manner for use by stakeholders, (which represent those involved in the industrial project in question).

The increasing use of compression of project execution times, in which the engineering phases of the various disciplines that make up the project must occur in parallel with each other, and in parallel with the construction and assembly phases, reducing the risk of even greater serious problems, with damaging impacts for the whole enterprise.

In this scenario, the Management of Interfaces between the different disciplines becomes a factor of great importance for the success of the enterprise through a correct compatibility and adequacy of the engineering documents. The purpose of this article is to discuss the importance of managing the interfaces and recommend a way to do this management appropriate to an ideal Maturity Model in Project Management. The focus will be the design phase of large industrial projects, consisting of documents such as flowcharts, data sheets, layouts, diagrams, typical details, among others.

For the development of the subject of this article, there is a review of the bibliography that shows some references of the literature on Interfaces Management and Maturity of Projects. It then addresses interfaces management in industrial projects and proposes two processes for this purpose.

After these analyzes and discussions, the article presents conclusions about the importance of Interfaces Management in industrial projects and how this can be done through a multidisciplinary coordination.

*Figure 1. Processes for interfaces management*
*Source: Prepared by the author himself*

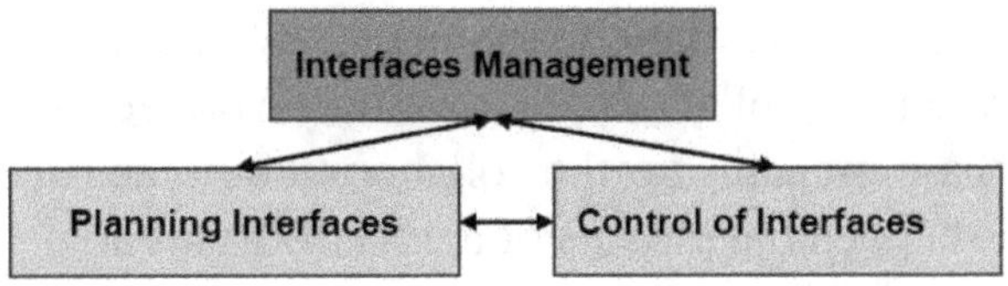

## DEVELOPMENT

### How Management of Interfaces Is Inserted in the Maturity of Complex Projects

The concept of interface management is defined in management of the physical and functional commitments of items that relate, ensuring compatibility between parts that adjust or interfere during the execution of the program / project and throughout the product or system life cycle, providing communication between the parties involved, physical and functional requirements and functional commitments between people, program sectors and organizations.

The focus for Interfaces Management (the aim of this article) is on the importance of multidisciplinary management in industrial projects and the compatibility of engineering disciplines. The following is a review of the literature. According to Casarotto (2010), large and well-managed companies usually have, in an organized way, the planning function or, more strictly, they have the expansion function, responsible for the coordination of studies and projects of their development and industrial growth. Projects are usually divided into the following phases: Feasibility, Conceptual, Basic Engineering, Detail Engineering, Assembly and Commissioning (commissioning).

For Maximiano (2002), effective administration of interfaces is one of the ways to increase the success probability of the project. Much of the quality of planning, organization, and other project functions depends much more on the articulation of agreements than on technical sophistication. According to Muto (2008), the planning of interfaces is responsible for identifying and mapping the interrelationships that exist within the program that belong to the organization's portfolio or to factors external to the program.

According to Ferreira (2007), for the project to come to an end efficiently, the need for coordination and collaboration is immense. Effort for collaboration is even more necessary and coordination is more difficult in engineering and construction ventures than in mass-production industries, precisely because it is not a mass production or assembly line, often equipment on order. In addition, collaboration at the design stage is essential so that all other steps can be well coordinated. Collaboration in other phases is necessary, but it is much more efficient when the collaboration starts on the project. The collaboration in the design phase prepares for the collaboration in the subsequent phases and thus makes happen the process of compatibilization.

According to CMMI (2006), many problems of product integration arise from unknown or uncontrolled aspects, both internal and external to the industrial enterprise. Effective management of the requirements, specifications, and design of product components helps ensure that the deployment of the interfaces will be complete and compatible. Interface management should start very early and include maintaining consistency of interfaces throughout the product lifecycle, as well as conflict resolution, drift and change. According to Graziano (2003) compatibilization is defined as an attribute of the project, whose components of the systems occupy spaces that do not conflict with each other and, in addition, the shared data have consistency and reliability until the end of the project and work process.

Analyzing the definitions of the Interfaces Management of the authors mentioned above, it becomes important to identify these concepts to identify an ideal Maturity model in Project Management. Thus, according to Cooke-Davies (2002), there are more than 30 maturity models currently on the market. These models are based on the concepts advocated by the areas of quality management (Crosby, 1979; Deming, 1986, as cited in Lukosevicius, Filho, & Costa, p 5, 2006).

Pennpacker and Grant (2003 as cited in Lukosevicius, Filho, Costa, p 6, 2006) say that "Maturity models provide a framework for evaluation that allows an organization to compare its project deliverables with the best practices or against its competitors," he says. defining a structured route for improvement "

Thus the practices and techniques of project management are recognized by many organizations of several industries as essential competencies that bring benefits to the business.

## Analysis and Discussion: Management of Interfaces in Industrial Projects

As was seen in the literature review, industrial enterprises are divided into several phases, which require efficient process management techniques and practices. This article is focused on the Project Engineering phase. This choice is due to the fact that the decisions taken at this stage, right or wrong, have a very large impact on the project and also because, normally at this stage, the project teams are physically separated, often even in different countries, with chance of mistakes even greater. The interfaces to be addressed are among the different Engineering knowledge disciplines that make up an industrial project.

The proposal is an interface map to be used in any industrial project and also the creation of a timeline with the precedence relations between the interface items. Next, two processes will be presented to be executed during the life cycle of a project, in order to allow an adequate management of the interfaces.

The first step is to define the interface map. It includes the following disciplines: Electrical Engineering, Automation, Civil, Mechanical and Utilities. The interface map is shown in Figure 2.

The interface map briefly describes the information that should be generated. It is interesting that, according to the work of the project teams, an "Interfaces Dictionary" is created to accompany it. This dictionary should indicate the minimum technical data for each interface item. In this way, avoiding that only after the information delivered, it is discovered that it is incomplete or incompatible.

In the sequence it is necessary to develop an interface schedule, which is also an addendum to be used in the projects. The timeline should show the interface items distributed in the project phases in which they are to be delivered and the precedence relationship between them. The next step is the proposition of two processes for the management of interfaces during the life cycle of each industrial project. These are: Planning of interfaces and Control of interfaces.

## Planning Interfaces

The planning of the interfaces should happen together with the planning of the scope, term, cost, etc., once you interfere you experience interference from these areas. Assuming the scope to be executed in the project and the map (and dictionary) of interfaces, the project team must define which interface items actually apply to the project. During this check it is possible that some interface items are not applicable and some others should be added. To register these interface items, the Interfaces Tracking Tab is suggested. This tab is a dynamic record of the status of project interface items. The proposed template for the same is similar to the map of interfaces, but with the inclusion of two columns: The indication of the responsible for generating the information regarding each interface item and the date of the updated information.

*Table 1. Map of interfaces for industrial projects*

| Map of Interfaces | | | | |
|---|---|---|---|---|
| **Interface Item** | **Who** | **Does what?** | **Who use** | **As?** |
| IA1 | Automation | List of automation and control signals | Electrical | Definition of I / O list |
| IA2 | Automation | Basic communication network architecture | Electrical | Define network cable route |
| IA3 | Electrical | Preliminary dimensions and weight of electrical equipment | Civil | Scale rooms and electrical buildings |
| IA4 | Electrical | Thermal Dissipation of Electrical Equipment | Civil | Dimension air conditioning system |
| IA5 | Electrical | Landfill Project and Electric Sausages | Civil | Base for civil detail of sausages |
| IA6 | Electrical | List of Remote PLC Inputs and Outputs | Automation | For the development of automation software |
| IA7 | Electrical | Basic routing of cables | Utility | Check interference with piping |
| IA8 | Electrical | Electrical system data | Mechanical Engineering | Proper specification of equipment |
| IA9 | Electrical | Base for oil transformers | Civil | Define oil collection well |
| IA10 | Metallic structures | Load Plan | Civil | Elaborate calculations |
| IA11 | Metallic structures | Lay out building | Mechanical Engineering | Locate equipment |
| IA12 | Mechanical Engineering | Plane of loads and weights of the equipment | Civil | Elaborate calculations |
| IA13 | Process | Discretionary of mechanical equipment | Automation | For the development of automation software |
| IA14 | Process | Lists of motor loads / electric loads (indicating power and type of drive) | Electrical / Automation | Dimension and elaborate distribution system of electrical energy |
| IA15 | Process | List of instruments and valves | Electrical / Automation | Elaborate automation system and specify instruments for purchase |
| IA16 | Process | Rental of plant equipment | Electrical | Elaborate route of electrical cables and divide PLC remote in the field |
| IA17 | Process | Flow of processes and instrumentation (P & I) | Automation | Elaborate list of signs |

Source: Prepared by the author himself

The monitoring interface interface items, although based on the interface map, can be grouped or divided according to the scope of the project, in order to facilitate the control of the interface interfaces. For example, you can subdivide an interface item into the subsystems and areas of the project or according to the specific responsibility of generating information. Once all the project-specific interface items are defined, they must be entered as activities in the project schedule. The precedence relations between the interface items and those with the other project activities should be indicated, in addition to the time needed to generate this information. The interface schedule will assist in performing this activity.

Once this data is defined, the project schedule must be re-checked since, with the inclusion of interface items, the project schedule may need to be modified, either through a delay of the project deadline or an increase in project resources of some activities in order to carry out the work with a shorter duration.

Like any activity in a project, it is important that the management of interfaces also has a responsible one. This person should coordinate the planning and, during project execution, should control the interfaces to ensure that the information is made available at the time and with the appropriate content. This assignment may be given to the project manager or to a person specifically designated for that purpose, according to the size and complexity of the project.

The Interfaces Planning process can be graphically represented for easy visualization, as shown in Figure 3. The planning of the interfaces should also determine the periodicity, the forms of communication and the people and companies involved in the process.

## Control of Interfaces

Once planned, the management of interfaces has to be executed and controlled. That is why this process is suggested.

The key point for controlling interfaces is communication. The person responsible for managing the interfaces is who will ensure such communication and manage any conflicts that may arise. The interface items should be controlled with the Interfaces Tracking Tab, by updating the columns: responsible and date of the updated information, preferably in specific meetings for this purpose.

*Figure 2. Interfaces planning process*
*Source: Prepared by the author himself*

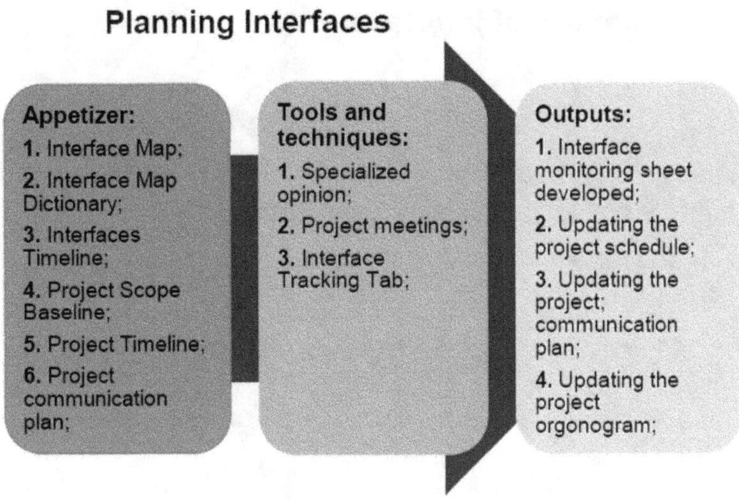


Interface problems are not often causes of project delays. Proper management will reduce the risk of these delays occurring by taking actions taken in advance. A graphical view of the Interfaces Control process is found in Figure 3 below.

The interface manager can avoid project quality and timing problems by making an appropriate follow-up of the interface items with the teams preparing the information. This monitoring is intended to ensure that the information is being delivered complete, as determined in the Interfaces Map Dictionary and on the appropriate date.

## Proposal for How to Choose Model of Maturity of Complex Engineering Projects From Interfaces Management

Following these three strategic steps defined within the Project Management Interfaces plan presented earlier, it becomes possible to identify the compatibility between the plan and the PMI (2003) Organizational Project Management Maturity Model (OPM3).

PMI (2003) asserts that the role of the model is to help organizations assess and understand their current level of organizational maturity by providing a path to improvement by recognizing which competencies of the organization should be developed. This directly reflects the organization's ability to enable its strategies through successful selection and completion

*Figure 3. Control of interfaces*
*Source: Prepared by the author himself*

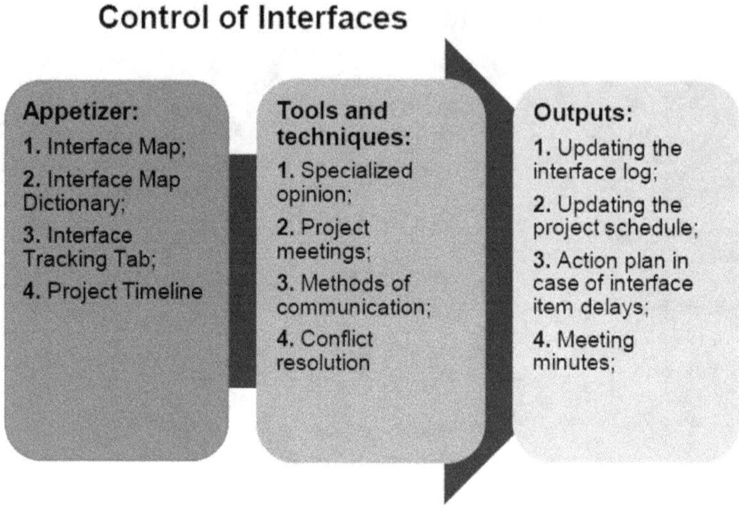


of its multiple projects, as well as to benchmark both internal and external (with other organizations) over time. The model is scalable and flexible, allowing its use in organizations of various industries and of various sizes.

OPM3 has several ways to classify the maturity of an organization's project management. This multi-visibility perspective provides the organization with more flexibility and better support in the decision-making process, as well as in the planning of improvements to be implemented. Thus, the model refers to: dimensions, domains, groups of project management processes and stages of process improvement.

OPM3 is organized in three interconnected dimensions (PMI, 2003):

- **Knowledge:** Explains the importance of project management at the organizational level and how organizational maturity in project management can be achieved.
- **Evaluation:** Shows the procedures that allow an organization to measure itself against the best practices described in the knowledge dimension.
- **Improvement:** It supports the selection and definition of alternatives that will lead the organization from the current state of maturity to a desired and more suitable state of maturity. This dimension is triggered in the model if the organization desires. Figure 4 shows the

*Figure 4. Cycle of application of OPM3*
*Source: Adapted from PMI (2003)*

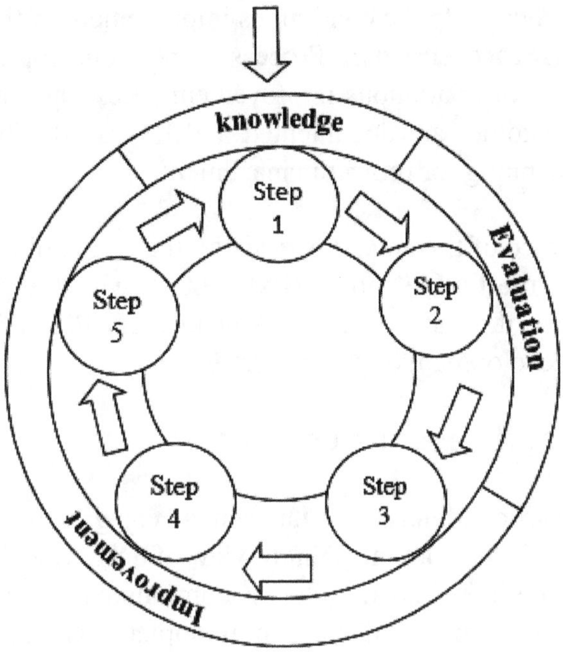


representative diagram of the OPM3 application cycle with the five steps of an organization's evaluation process.

With this, the Interfaces Management fits in the five steps of the OPM3 model.

- **Step 1 (Knowledge):** Preparation for evaluation (introduction to project management or common language.
- **Step 2 (Evaluation):** Execution of the evaluation (common processes): The needs to implement processes and methodologies that support project management are perceived. It is recognized the importance of acquiring knowledge in project management. Project management is done for the scope, time, and cost.
- **Step 3 (Improvement):** Improvement planning (single methodology or standardization of project management): processes are integrated and there is support and recognition of top management.

- **Step 4 (Improvement):** implementation of the improvements (Benchmarking process): a project management office is mandatory and it is dedicated to the continuous improvement of the processes.
- **Step 5 (Improvement):** Process repetition (optimized project management or continuous improvement): Lessons learned from one project to another are documented and leveraged. There is ongoing strategic planning for project management.

Thus, the choice of this model is made by the compatibility between the requirements of the OPM3 model (PMI 2003), with the provenance and reliability of the information that the Management Plan of Interfaces can provide in the development of an enterprise.

## Conclusion and Lessons Learned

As noted in this article, Interfaces Management can greatly contribute to the success of the project. From the point of view of quality, with an integration between the disciplines and the closest communication in the meetings, interface information can be delivered more completely and correctly, avoiding rework. Interface management can also help projects to meet planned deadlines, since interfaces that are often overlooked or not properly handled mean delays in the projects. Proper planning and frequent and effective monitoring of interface information can ensure that it is delivered at the right time.

Communication is the key to ensuring that information is properly distributed, in the right review, to the right people. In this way, responsibility for managing interfaces should be assigned to a specific person and meetings for checking interface items should be frequent. The recommendation presented, although specific to the engineering phase of industrial projects, can also be applied in other phases of industrial projects. For this, it is enough that the interface map, the schedule and the interface monitoring sheet are elaborated with the specific information, aiming at the compatibility of other steps (eg, supplies, manufacturing, construction, assembly). For the benefits presented, it is the reflection that the effort and time applied to the Management of Interfaces, as opposed to costs, should be considered as an investment that will certainly avoid a series of delays, rework and modifications of equipment due to the problems arising from the mismanagement of the information, thus contributing to the success of the enterprise.

## CONCLUSION

Developing an effective Interfaces Management Plan has proven to be feasible for choosing an ideal Maturity Model in Project Management and Construction. Good planning and application of the stages such as Map, Planning and Interfaces Control in the development of a large enterprise can guarantee the fulfillment of deadlines and allocation of resources, fostering transparency in the relations and needs of stakeholders.

The Interfaces Management Plan showed how much it can contribute to the communication of the projects and subprojects involved in a project, allowing a greater reliability of the information of each phase, building a traceability system when there are requests for information by the parties involved in the project.

## REFERENCES

CMMI. Product Team. (2006). *CMMI for Development, Version 1.2*. SEI. Disponivél em: https://www.sei.cmu.edu/reports/06tr008.pdf

Ferreira, S. L. (2007). Da engenharia simultânea ao modelo de informações de construção (BIM): Contribuição das ferramentas ao processo de projeto e produção e vice-versa. *VII Workshop Brasileiro de gestão do processo de projetos na construção de edifícios.*

Filho, C. N. (2010). Elaboração de projetos empresariais: análise estratégica, estudo de viabilidade e plano de negócio. São Paulo: Atlas.

Graziano, F. P. (2003). *Compatibilização de Projetos.* Instituto de Pesquisa Tecnológica – IPT. Disponível em: https://www.revistas.usp.br/gestaodeprojetos/article/download/50928/55010

Lukosevicius, A. P., Filho, L. A. N. C., & Costa, H. G. (2006). *Maturidade Em Gerenciamento De Projetos E Desempenho Dos Projetos, 2006.* Disponível em: www.producao.uff.br/conteudo/rpep/volume72007/RelPesq_V7_2007_07.doc

Maximiano, A. C. A. (2002). *Administração de projetos: Como transformar idéias em resultados. 2.* São Paulo: Editora Atlas.

Muto, C. A. (2008). *Guia para gestão de programa e múltiplos projetos do conceito à prática.* Disponível em: www.brasport.com.br

# Chapter 8
# Stakeholders in Engineering and Construction Projects


## ABSTRACT

*This chapter describes what stakeholders are, how to classify them, how they can affect and be affected by the projects, and then describes how to establish a dialogue with them to achieve success in project implementation. Throughout this chapter a series of specific observations are made for complex projects with the purpose of contextualizing the subject of the chapter for this type of project. What is the relevance of stakeholders for complex projects?*


## INTRODUCTION

What is the relevance of stakeholders for complex projects? In the third chapter of this book we highlight three characteristics of a complex project:

- Multiple stakeholders
- Ambiguities
- Multinationality

In that chapter we mentioned that different stakeholders may have different expectations, which can compete with each other and create conflicts within the project. We also emphasize that ambiguities and multinationalities have, by themselves, strong relationships with stakeholders regardless of the project being complex or not.

DOI: 10.4018/978-1-5225-5864-4.ch008

In the case of ambiguities, different stakeholders may have ambiguous positions regarding a project, which add to the ambiguities of other natures that this project also has.

Taking back an example previously used, many mining projects or hydroelectric power plant projects are characterized by having ambiguous repositions stakeholders. These projects have stakeholders directly related to the enterprise and also indirectly linked. In this second group we highlight:

- Public administration
- Regulatory agencies
- Non-governmental organizations (NGOs)
- Local merchants
- Local workers
- Residents of the community

Some of the ambiguities that may be associated with them are:

- Will public administration and regulatory agencies provide the necessary (environmental and other) licenses for the project to be deployed and operated?
- What is the position of the residents in relation to the project? Local merchants and workers can see some benefit in the implementation of the project, but what about the others?
- Can the conditions imposed by the public administration and regulatory agencies, based on the demands of the community, for the issuance of licenses be able to derail the project or to reduce its profitability significantly?
- Can events occur during the implementation of the project that stresses the relationship between the proprietary organization and the community?

Regarding multinationality, which should be understood as the characteristic of projects whose activities are dispersed in different parts of the globe, we believe that it can make the question of stakeholders more complex, since it adds the ingredient of potentially significant cultural differences between them. These cultural differences can bring new ambiguities to the project.

So, projects (or organizations) must incorporate in their strategies the interests and aspirations of all the stakeholders that directly or indirectly relate to them. Thus, it is necessary to map, identify, analyze and classify these stakeholders, so that they can be included in the organizations, global strategy and can be managed as business and project components.

Cleland and Ireland (as cited in Costa, Campos Filho and Lukosevicius, 2007) relate some factors that evidence the success of a project, such as: the project objectives were met within the deadline and budget, team members consider participation in the project a valuable experience, the result of the project generated profit or some competitive advantage for the organization.

However, the same authors raise a question regarding the relationship between stakeholders and the success of a project. According to them the success or failure of a project can be perceived differently by stakeholders. This exposes the evaluation of the performance of the project to the subjectivity of the gaze of those who evaluate it.

Regardless of how these positions can be reconciled, the fact is that stakeholders have a relevant role in the development and success (or failure) of complex projects.

## STAKEHOLDERS

The term stakeholder arose in 1963, in a consulting service made by the Stanford Research Institute, as a way of describing a group of people and organizations that are not shareholders, but without their support the company may cease to operate or exist (Silva Neto, 2011).

The Project Management Institute (PMI, 2013) defines stakeholders and some of their characteristics as listed below:

- Stakeholders are people, groups, or organizations that could be impacted by the project.
- They are comprised of persons and organizations such as customers, sponsors, the performing organization, and the public (who are involved in the project or whose interests may be affected by the project).
- Every project will have stakeholders who are impacted by or can impact the project in a positive or negative way.
- Most projects will have a diverse number of stakeholders depending on their size, type, and complexity.
- They may also influence the project and its deliverables.

- The ability of stakeholders to influence the project is typically highest during the initial stages and progressively gets lower as the project progresses.
- Stakeholder satisfaction should be managed as a key project objective.

Although shareholders were the first and most constant type of stakeholder of a company identified by management theory, it was with the broadening of the understanding about who is and who composes the stakeholders that the theme could be inserted in the management agenda and methodologies of projects.

Originally the organizational performance was considered basically under the profit generating bias, which means that the shareholders were considered in the first place, being the other stakeholders considered to the extent that they could contribute to the generation of these profits.

In the last decades a new understanding has emerged, which considers that the objective of the organizations is to take care of the interests of all the stakeholders, having no more or less legitimate interests. Add to this the understanding that any stakeholder can influence the profit generation of an organization, and thus they become of great interest to it.

## TYPES OF STAKEHOLDERS

There are several classifications used for stakeholders. Let's cite some of them here, which are fully compatible with each other.

According to the PMI (2013) stakeholders can be internal or external. Stakeholders may be at different levels within the organization and may possess different authority levels. They are the internals stakeholders. Other kind of stakeholders is external to the performing organization for the project. They are the external ones.

Brito and Terra (as cited in Moura, 2013) classifies stakeholders as direct and indirect, as follows:

*[Stakeholders] are interested parties or affected parties in any subject not necessarily economic. Their relationship with the company or organization is often indirect and does not necessarily involve business transactions. Typically, stakeholders are divided into a more direct group (such as employees and shareholders, financial institutions, suppliers and clients) and a more indirect group (such as communities, government, media, interest groups, competition*

*and advocacy groups). Stakeholders may stand behind an advantage, a repair or an insertion in a larger context.*

An analysis of this description allows us to conclude that the direct and indirect stakeholders of Brito and Terra correspond, respectively, to the definition of internal and external stakeholders that is used by PMI.

Kerzner (2006) proposes a project stakeholders structure, as shown below, which does not conflict with previous definitions, but being presented in a more detailed way will certainly be useful in some of the Stakeholder Management processes.

*Figure 1. Interested in the project*
*Source: Kerzner (2006)*

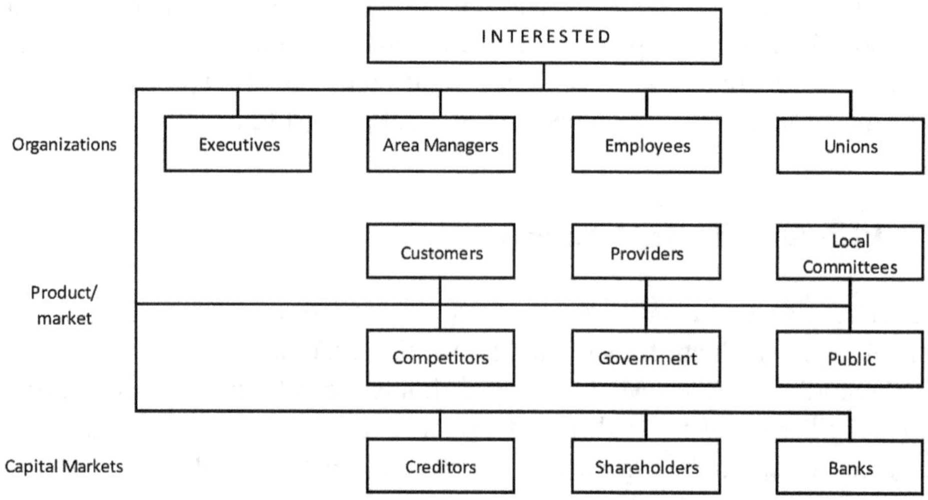


*Table 1. Stakeholders engagement criteria*

| Engagement | Description |
|---|---|
| Unaware | Unaware of project and potential impacts. |
| Resistant | Aware of project and potential impacts and resistant to change. |
| Neutral | Aware of project yet neither supportive nor resistant. |
| Supportive | Aware of project and potential impacts and supportive to change. |
| Leading | Aware of project and potential impacts and actively engaged in ensuring the project is a success. |

Source: Author based on PMI (2013)

The PMI (2013) also classifies the stakeholders according to their engagement with the project. This classification is shown in Table 1.

However as varied groups (or types) of stakeholders as can be, it is important to note that an individual can incorporate many of these types at the same time. In other words, an individual could be a citizen, a shareholder and an employee of a company, inserting it in a multi stakeholder perspective.

Another aspect to be considered is the network of relationships between stakeholders of different groups. This may be especially relevant in small communities where there are many kinship or friendship relationships between individuals from different groups. For example: The mayor (public administration) of a small community may be the brother of the main local merchant who aspires to be a future supplier of the project being deployed.

In other words, we should not see stakeholder groups as extremely rigid, on the contrary, we must bear in mind that there are intersections between them.

In this chapter we will use the classification of internal and external stakeholders, with special focus on external stakeholders.

## HOW STAKEHOLDERS CAN AFFECT A PROJECT

According to the PMI (2013), some stakeholders may have limited ability to influence the project, others may have significant influence on the project and its expected outcomes. So, some groups of stakeholders (external) can bring significant losses to projects in their various aspects.

Stakeholder action can lead to significant changes in the project's scope and schedule, with consequent repercussions on its budget as well. In addition, the wear and tear on the relationship between the proprietary organization and these groups can become a source of future losses for the organization.

According to Peixoto, Silva Neto and Mazzuchelli (2010), in the last decade there have been few cases of projects that have been delayed or company operations paralyzed by actions of stakeholders such as organized communities, public administration, regulatory agencies, the press and environmental organizations, among others. To the extent that a postponed project or a paralyzed operation has a direct impact on the generation of resources by the company, it is clear the extent to which the stakeholder action can have on it.

In order to highlight this relationship between stakeholders and losses, Moura (2013) related the expressions used by business magazines to describe the cause of the significant losses that occurred in four mining projects of Brazilian companies. The articles in these journals use expressions such as:

- Access to land;
- Political articulation;
- Endless disputes in justice;
- Requirement of new licenses;
- Political instability;
- Interruption of mining concessions;
- Environmental and other licenses;
- Environmental bodies;
- Suspects of corruption.

The above expressions are directly associated with stakeholders, such as public administration, regulatory agencies, communities directly affected by projects, organized society and NGOs among others, evidencing the relationship between stakeholders and losses.

One situation that deserves to be highlighted is when the one organization of one project is replaced by another, that is, the project was sold to another organization. The stakeholders' perceptions regarding the new organization may be different, causing them to change their positioning relating to risk by changing the likelihood and/or impact of the risk.

In other words, changing the ownership of a project will not change the risks themselves, but it can change their likelihood and/or impact as a result of stakeholder repositioning.

This naturally leads us to the next item in this chapter, which is Stakeholder Management. One of the challenges in project management, especially complex ones, is to avoid or minimize the losses that can be caused by an improper relationship with stakeholders. According to PMI (2013), the ability of the project manager to correctly identify and manage these stakeholders in an appropriate manner can mean the difference between success and failure.

## STAKEHOLDERS MANAGEMENT

In 2003 Grayson and Hodges proposed seven steps to minimize risk and maximize stakeholder opportunities, as follows:

1. Identify the triggers;
2. Prepare solid arguments;
3. Evaluate the scope of the issues;
4. Commit to action;

5. Integrate strategies;
6. Involve stakeholders;
7. Do the evaluation and the reports.

The description of these steps includes, among others, the following actions:

- Open dialogue with stakeholders;
- Assess community needs;
- Know the risks;
- Manage risks.

Although the approach proposed by Grayson and Hodges (2003) does not have the processes structure of other approaches, it has the merit of establishing that for organizations, there is a relationship between stakeholders and risks.

According to the Institute of Risk Management (2003), the benefits and opportunities should be viewed not just in the context of the activity itself but in relation to the many and varied stakeholders who can be affected.

Standard NBR 31000 Risk Management - Principles and Guidelines, Brazilian version of ISO 31000: 2009, defines that:

*All the activities of an organization involve risk. Organizations manage risk by identifying, analyzing, and then assessing whether risk should be modified by risk management in order to meet their risk criteria. Throughout this process, they communicate and consult stakeholders, monitor and critically analyze the risk and the controls that modify it to ensure that no additional risk treatment is required.*

According to the PMI (2013):

*Stakeholder management is about creation and maintenance of relationships between the project team and stakeholders, with the aim to satisfy their respective needs and requirements within project boundaries... focuses on continuous communication with stakeholders to understand their needs and expectations, addressing issues as they occur, managing conflicting interests and fostering appropriate stakeholder engagement in project decisions and activities.*

*Table 2. Process of stakeholders management*

| Process | Description |
|---|---|
| Identify Stakeholders | The process of identifying people, groups, or organizations that could impact or be impacted by a decision, activity, or outcome of the project; and analyzing and documenting relevant information regarding their interests, involvement, interdependencies, influence, and potential impact on project success. |
| Plan Stakeholder Engagement | The process of developing appropriate management strategies to effectively engage stakeholders throughout the project life cycle, based on the analysis of their needs, interests, and potential impact on project success. |
| Manage Stakeholder Engagement | The process of communicating and working with stakeholders to meet their needs/ expectations, address issues as they occur, and foster appropriate stakeholder engagement in project activities throughout the project life cycle. |
| Monitor Stakeholder Engagement | The process of monitoring overall project stakeholder relationships and adjusting strategies and plans for engaging stakeholders. |

Source: Author based on PMI (2013) and Boyadjian (2017)

The sixth edition of the PMI Project Management Book of Knowledge (PMBOK) (as cited by Boyadjian, 2017) structures stakeholder management in the fourth process as described in Table 2.

It is important to highlight that in the PMBOK the Stakeholder Management knowledge area has a very strong relationship with other areas, especially Risk Management, Communication Management and Resource Management. In other words, talking about Stakeholder Management also involves concepts and processes from these three other areas.

Resource Management, in the sixth edition of the PMBOK, replaces the Human Resources Management process of the fifth edition.

Managing stakeholders is above all dialogue with stakeholders. Dialogue, in the sense that it is necessary a mutual knowledge, that company and society understand their respective demands and offers, that society understands the moment of the company and vice versa.

In this context, an important consideration is when to initiate management and consequently dialogue with stakeholders.

According to the PMBOK (2013) formal identification of all stakeholders should be introduced as a component of the project and conducted from the outset.

Silva Neto (2011) in a study on mining projects in Brazil observed a different practice. It was observed that rarely projects contemplated the management and dialogue with the stakeholders from the beginning. In many cases they only started the dialogue after the start of the environmental licensing process; reaching extreme situations of identifying companies dialoguing with the community only at the time of so-called public hearings, mandatory legal framework for environmental licensing in Brazil.

Stakeholder dialogue should not and cannot have a roadmap linked only to legal requirements and cannot walk "in tow" of environmental licensing. It is necessary that it be carried out with its own procedures and itineraries, gaining the body of a social license to operate (SLTO). It should begin along with the project management process, or in the pre-project stage itself. The fact is that it must begin at a stage prior to environmental licensing.

In a pre-project phase, it should be explicit in the dialogue that the fact that the company is related does not mean that there is certainty that the project will be viable, but rather that it has a correct understanding of what it means to integrate a social environment and that it will maintain a dialogue with its interlocutors, informing about the context and the scenario of the project, about advances and setbacks.

Although the above observations are based on a study of specific scope (mining projects in Brazil) we understand that the guidelines and conclusions of Silva Neto are also valid for a diversity of other projects in other locations.

It is clear that, starting from the pre-project stage, stakeholder management will extend through all phases of the project until its closure.

We will use for the development of this chapter the processes proposed by PMI focused on external stakeholders.

## Identify Stakeholders

According to the PMI (2013), identify stakeholders early in the project is critical for its success. Identify stakeholders covers analyze their levels of interest, their individual expectations, as well as their importance and influence. This initial assessment should be reviewed and updated regularly. It is made analyzing and documenting relevant information regarding their interests, involvement, interdependencies, influence, and potential impact on project success.

The PMI proposes some criteria for classifying stakeholders to be used to identify and group them. Some of these criteria are described in Table 3.

Another important aspect of stakeholder identification is the identification of the relationship networks that exist among stakeholders, since stakeholder groups such as these are not rigid and isolated. On the contrary, we must bear in mind that there are intersections between them. We must then identify

*Table 3. Stakeholders classification criteria*

| Classification Criteria | Description |
|---|---|
| Power/ influence grid | Grouping the stakeholders based on their level of authority ("power") and their active involvement ("influence") in the project. |
| Influence/ impact grid | Grouping the stakeholders based on their active involvement ("influence") in the project and their ability to effect changes to the project's planning or execution ("impact"). |
| Salience model | Describing classes of stakeholders based on their power (ability to impose their will), urgency (need for immediate attention), and legitimacy (their involvement is appropriate). |

Source: Author based on PMI (2013).

these networks of relationships that will allow us to find several points of insertion between them.

## Plan Stakeholders Engagement

In this process the main product will be the stakeholder management plan. It may be formal or informal, highly detailed or not, depending on the needs of the project.

In order to elaborate this plan, adequate strategies must be developed to effectively involve stakeholders throughout the life cycle of the project, based on the analysis of their needs, interests and possible impacts on the success of the project.

Of particular importance will be the database of lessons learned and historical information as they provide information about previous stakeholder management plans and their effectiveness.

The plan will also cover several communication concepts, such as:

- Stakeholder communication requirements for the current project phase;
- Information to be distributed to stakeholders, including language, format, content, and level of detail;
- Reason for the distribution of that information and the expected impact to stakeholder engagement;
- Time frame and frequency for the distribution of required information to stakeholders;

- Method for updating and refining the stakeholder management plan as the project progresses and develops.

## Manage Stakeholders Engagement

According to PMI (2013), Manage Stakeholder Engagement is the process of communicating and working with stakeholders to meet their needs/ expectations, address issues as they occur, and foster appropriate stakeholder engagement in project activities throughout the project life cycle.

Special attention should be given to the continuity of the dialogue process with stakeholders.

It is observed that companies, either with their own or outsourced teams, invest in the mapping process and arrive well to the point where they design their relationship matrix, classifying their stakeholders by power, influence, legitimacy, as well as defining the communication strategy to be adopted as the dialogue goes deeper. However, from this moment on, there is usually a slowdown or, in some cases, the abandonment of the process.

As causes, two situations are clearly identified, both related to a transition process. In cases where the work has been carried out by third parties up to this point, it is common that, at this moment, in which the dialogue must be carried out, as a priority, by the company's own actors, there is a difficulty of the team, even though it has been involved during the whole process, to take their driving. In addition, moving from planning to practice requires companies to involve not one or two professionals, but rather a team composed of the most diverse internal stakeholders who, not always, take up positions or have the qualifications required for dialogue with the plurality of external stakeholders.

The most common is that the process ends up being in charge of professionals related to the areas of communication or social responsibility, which causes a deceleration and depletion of the process.

## Monitor Stakeholders Management

Monitoring stakeholder involvement is the process of monitoring stakeholder positioning in relation to the project and the relationships among the various stakeholders, among other aspects.

Due to monitoring, adjustments to project strategies and plans may be necessary. Some of the plans that may require updating are as follows:

- Change management plan;
- Communications management plan;
- Cost management plan;
- Human resources management plan;
- Procurement management plan;
- Quality management plan;
- Requirements management plan;
- Risk management plan;
- Schedule management plan;
- Scope management plan;
- Stakeholder management plan.

## STAKEHOLDERS EXTERNOS

Annually, two consultancies (Deloitte and EY) issue reports on the risks associated with the mining sector. This sector is characterized by typically complex projects (multiple stakeholders, ambiguities and multinationalities). Thus, some of the points raised by these two consultancies are certainly applicable to complex projects in other sectors.

The 2016-2017 EY report (2016) lists the top 10 risks to the mining sector as follows:

1. Cash optimization
2. Capital Access
3. Productivity
4. Social License to Operate (SLTO)
5. Transparency
6. Switch to grow
7. Access to energy
8. Joint ventures
9. Cybersecurity
10. Innovation

Analyzing the relationship above, three risks that have a direct relationship with external stakeholders stand out. These are: Social License to Operate (SLTO), transparency and access to energy. We will use these risks to discuss relevant aspects in the relationship of organizations and their projects with external stakeholders.

## Social License to Operate

SLTO can be defined as the agreement of external stakeholders regarding the existence of the project. In this context, EY (2016) states:

*To maintain a strong SLTO, it is important to integrate sustainability into long-term planning and link key performance indicators with productivity outcomes as well as remuneration structures. Successful companies have highlighted the value of operating in tandem with communities and have shown that it is possible to engage through mutual value creation — for example, offering some local ownership and sourcing labor and procurement domestically will provide a greater level of community engagement.*

Safety, health and environment issues are important not only for meeting internal stakeholders and legal requirements, but mainly for maintaining SLTO.

The local community includes art from project workers. The same community shares the environment with the project. So it is natural for the community to position itself critically in relation to the owner organization or its project if it understands that safety, health or the environment are not being adequately managed.

Another aspect related to SLTO raised by Deloitte (2015) is on how to evaluate the performance and evolutions in the relationship between the organization and the community. It suggests the use of performance indicators in the dialogue with the community, which should have relevant meaning for the community. Here are some examples suggested by the consultancy:

- The growth and profitability of local suppliers;
- The training opportunities made available to employees;
- Improvements in literacy and secondary school completion rates;
- The percentage of children completing traditional or spiritual education programs;
- Improvements in softer measures, such as work ethic or employee satisfaction.

Here again we see the fact that there are intersections between different groups of stakeholders, employees and community especially in this item.

## Transparency

The owner organization must pay attention to transparency in the relationship with the various stakeholders. One of the objectives of this is to avoid that the dialogue with them and their image is compromised by the conflict of interest between different stakeholders. It is understood conflict of interests with the situation in which different stakeholders have conflicting interests and expectations regarding the project.

An example: Are local government interests in relation to a given project the same as in the local community? If not, how should the organization deal with this so as not to have the dialogue established with them and their image compromised by this conflict of interests. The way the organization deals with this starts with transparency with both stakeholders.

Obviously, we are rejecting here any illicit "solution" such as bribery or other corrupt practices.

Improving transparency can help organizations show their social contribution by providing local communities with greater access to information and, in turn, helping organizations to refute claims that they are not fulfilling their role.

An example of transparency initiatives are industry organizations created to promote it, such as the Extractive Industries Transparency Initiative (EITI) created in 2003. This was followed by many other country and region-specific outreach initiatives in the US, Canada and Europe.

Another example is the Rio Tinto mining company. Since 2010, it publishes annual report with all taxes paid. This initiative was followed by other large global mining companies, which began to voluntarily disclose their government payments.

## Access to Energy, Water and Land

The risk related to access to energy, and other natural resources, is when a dispute is established between the project and the local community regarding the priority of using these locally scarce resources.

For industrial plant projects whose energy consumption is high (some mining projects and steel projects, for example) an important consideration when choosing an appropriate source of energy. However, this is only one aspect of a much greater strategic decision. Other considerations are:

- Uncertainty of availability and cost of energy over the entire project life;
- Counterparty risks;
- Operational alternatives;
- Social, financial and reputational implications of every choice.

Another aspect to be considered in the energy source to be chosen for the project is the social and imaging benefits that can be obtained by reducing the emission of greenhouse gases related to diesel.

Decisions regarding water source and land use have the same strategic bias as the choice of energy source and may have a similar impact on the relationship with external stakeholders.

According to Deloitte (2015):

*One of the factors that often spur tension between mining companies and local stakeholders revolves around competition for scarce resources. In many mining regions, there is insufficient power to meet the needs of both industry and local citizens. Similarly, land rights are coming into dispute, with local communities loath to relocate or give access to traditional hunting grounds. The same is true of water. In some regions, governments have put mining projects on hold in light of condemns around the impact they were having on the quality and availability of local water resources. In other regions, governments are trying to limit water rights previously granted to mining companies.*

## What to Do

In this topic we list some tips on how to relate to external stakeholders. They do not replace the processes defined by PMI for stakeholder management but add to them as they instrumentalise some of these processes or provide insights for organizations.

Deloitte (2015) proposes six actions for a correct relationship with external stakeholders in the mining sector. In Table 4 we can see a summary of the actions proposed without the specific focus in the mining sector.

The authors Silva Neto, Peixoto and Mazzuchelli (2010) also make an important contribution to the subject. They relate three weaknesses in how the relationship with stakeholders is conducted by mining companies in Brazil. They are:

*Table 4. Actions to build consensus*

| Action | Description |
| --- | --- |
| Get serious about social | It's time for companies to get more active in social media and use them to engage directly and share targeted information with their various stakeholders. |
| Listen carefully | Beyond using social media platforms to communicate with stakeholders, miners should also leverage data analytics and 'social listening' tools to track what is being said about their organizations in real-time. Social listening provides companies with an early warning system that allows them to respond proactively. |
| Demonstrate commitment | To create win-win platforms that align companies with the complex web of stakeholders, these companies must demonstrate a high Level of commitment by engaging senior executives—right up to the CEO—to play key roles in stakeholder engagement and solution identification. |
| Expand the dialogue | By collaborating with employees, suppliers, government ministries, citizen groups, advocacy organizations, and even secondary and tertiary businesses, companies can expand the set of potential compromise outcomes that might meet the needs of typically divergent local groups and co-op the entire universe of stakeholders into the solution-seeking process. |
| Help inform national mine strategies | Governments eager to attract mining investment without alienating either corporate or community citizens would likely welcome input from industry stakeholders as they work to structure effective national mine strategies. Mining companies have a role to play in this regard as governments seek direction around policy. |
| Walk away | Unreasonable pressures exerted by regulators or other stakeholders could tip an otherwise feasible project into the red. Companies capable of responsibly walking away from projects that no longer promise to deliver a solid business benefit would send a strong message to governments and local communities on what they potentially stand to lose by adopting an intransigent anti-project stance. |

Source: Author based on Deliotte (2015).

- Postpone the start of the Stakeholder Management process, identifying them late;
- Disregard the existing relationship networks between stakeholders from different groups;
- Discontinue the relationship with stakeholders, within the Stakeholder Management and Communication Management processes.

On networks of relationship, it is necessary to observe not only the actors that compose the groups, but also to identify their relationship networks, which will allow to find several insertion points between them. This care has a high relevance for the dialogue process that we establish with the actors, because as we define communication strategies with each group, knowing that there are areas of intersection between them, may mean that different approaches are necessary for the actors that are located there.

Regarding discontinuity in the relationship, a deceleration or even abandonment of the dialogue is commonly observed when it is primarily driven by people from the company itself. This is because these people, even having been involved throughout the process, have difficulties going from planning to practicing dialogue. Taking the dialogue requires companies to involve an entire team, not one or two professionals. This team should be formed by several internal stakeholders who, not always, effectively assume this role.

From what has been exposed in this topic it is evident that the process of management of external stakeholders has a strong relation with the processes of risk management and communication.

## INTERNAL STAKEHOLDERS

Although our focus is on external stakeholders, here are some quick considerations on internal stakeholders.

Remember that the process of External Stakeholder Management has a strong relationship with the Risk Management and Communication Management processes.

For internal stakeholders, the relationship is strong with the processes of Resource Management and Communication Management. The Resource Management process, in the sixth edition of the PMBOK, replaced the Human Resources Management process of the fifth edition.

The three tables below summarize the five maturity levels of three models: the CMM (Capability Maturity Model), the MMGP (Maturity Model in Project Management) and the PMMM (Project Management Maturity Model). These tables are our starting point for talking about internal stakeholders.

*Table 5. CMM Maturity Levels*

| Level | Name | Description |
|---|---|---|
| 1 | Initial | The process is disorganized and informal. Projects usually exceed the original deadlines and costs. Success depends on individual efforts and heroism. |
| 2 | Repetitive | Basic project management processes (functionality, schedule, and costs) are established. There is the discipline necessary to achieve success in projects similar to previous ones. |
| 3 | Defined | The standard of processes for software development exists. Management and engineering activities are documented, standardized, and integrated. All projects use these processes. |
| 4 | Managed | Quantitative control is established. Detailed measurements of process and product quality are carried out. That is, both the process and the product are understood and controlled quantitatively. |
| 5 | Optimized | The maturity model is institutionalized. Continuous improvement occurs through feedback of the process and the inclusion of innovative ideas and technologies. |

Source: Author.

*Table 6. Maturity levels of the MMGP model*

| Level | Name | Description |
|---|---|---|
| 1 | Initial, embryonnaire or *ad hoc* | The organization has not yet made any effort to implement a project management model. Projects are executed on the basis of intuition, goodwill and individual effort. Success is the fruit of individual effort or chance. |
| 2 | Known | The organization has awakened to the subject of project management. An effort was made to create a language for the subject of project management. This effort was made from different types of training. |
| 3 | Standardized | The organization has deployed and uses a standardized project management model based on competencies, processes and tools. A management platform was deployed. |
| 4 | Managed | The organization has consolidated the efficient and effective practice of the model implemented in Level 3 (organizational structure, methodology, computerization and strategic alignment). The organization uses really competent professionals. It also consolidates a strong alignment of projects with company strategies. |
| 5 | Optimized | The organization has attained the wisdom of project management. All initiatives implemented at previous levels have reached a level of excellence. The organization practices continuous improvement and technological and process innovation. |

Source: Author based on Prado (2015)

*Table 7. PMMM maturity levels*

| Level | Name | Description |
|---|---|---|
| 1 | Common language | The organization recognizes the importance of project management and the need to have a knowledge base to at least be able to establish terminology. |
| 2 | Common processes | The organization recognizes that common processes need to be defined and developed so that the success of one project can be replicated to everyone else in it. |
| 3 | Singular methodology | The organization recognizes the synergy generated by the combination of all corporate methodologies in a single one, whose central axis is project management. |
| 4 | Benchmarking | The organization recognizes that process improvement is necessary to maintain a competitive advantage. Benchmarking should be done on an ongoing basis. |
| 5 | Continuous improvement | The organization evaluates the information obtained from benchmarking and decides whether they can improve their methodology in project management. |

Source: Author

Analyzing the tables above, we can observe the following points in common between the models:

- Creation of project management standards;
- Disseminated use of project management standards in the organization;
- Use of benchmarking and continuous improvement process;

- Clearly establishing a relationship between project management, specifically of portfolios and the organization's strategy.

Implicit in the models is the need to:

- Disclosure to internal stakeholders;
- Training of them;
- Obtaining their understanding;
- Obtaining their agreement.

Thus, issues related to (human) resources and communication are essential for meeting these needs. This should be considered in resource and communication planning.

We must ensure that the public of these human resources and communication actions embrace (in addition to the employees of the organization) third parties, service providers, suppliers and clients, among other internal or direct stakeholders to the project.

Another aspect to be considered in multinational projects is the use of modern communication technologies to implement these actions.

## CONCLUSION

It is evident the need of Stakeholder Management, especially for the external ones, for a project to be successful. This management is a continuous dialogue between the organization and the various stakeholders, aiming at an alignment of expectations between the parties.

Complex projects are characterized by multiple stakeholders, ambiguities and multinationalities. These three aspects (stakeholders, ambiguities and multinationality) have a relationship that results in increased risks associated with stakeholders. Thus, for complex projects the relevance of stakeholders is even greater.

Stakeholder Management is obviously not a process that walks alone but is instrumented and instrumentalizes processes such as Risk Management, Communications Management and Resource Management.

Stakeholders influence the likelihood of a risk occurring and the potential impact of it. The positioning of stakeholders in relation to a project may change due to the actions of the organization of other factors. These repositioning can change the likelihood and/or impact of the risk in which these stakeholders are associated.

The beginning of the Stakeholder Management process must occur in the initial phases of the project. Thus, owner organization will have more time to develop the "dialogue" with the stakeholders and consequently they will have more time to make the right association between risk and stakeholders.

For certain types of projects, the issues related to energy source, water source and land use are relevant in relation to the local community due to local scarcity of theses resources.

In the identification of stakeholders, the network of relationships that exists between people of different groups must be taken into account, that is, the groups of stakeholders have intersections. In addition, the Stakeholder Management process should not be discontinued.

Another interesting aspect is that the owner organization of a project is changed, the probability and the impact of the risks can undergo changes due to the repositioning of the stakeholders. Thus, the reaction of the stakeholders should be observed when this exchange takes place. For example: The stakeholder relationship of a small town with a small national mining company may be different than with a large foreign company which bought the first one. This can lead to repositioning of stakeholders by changing the likelihood and impact of the risks.

Finally, it is important to highlight the important role of project managers in conducting this process. It is up to them to seek the convergence of interests and actions of internal stakeholders to the project so that the team, together, can also ensure that external groups are properly involved and included in the process. In this way, we will arrive at a good equation for the success of the project.

## REFERENCES

Boyadjian, J. C. (2017). *Comparativo do guia de gestão de projetos PMBOK® 2017 x 2013* [Comparison of the PMBOK® 2017 x 2013]. Retrieved from https://jcboyadjian.files.wordpress.com/2017/09/comparativo-pmbok-2013-x-20171.pdf

Costa, H. G., Campos Filho, L. A. N., & Lukosevicius, A. (2007). Maturidade em gerenciamento de projetos e desempenho dos projetos [Maturity in project management and project performance]. *Relatórios de Pesquisa em Engenharia de Produção, 7*(7).

Deloitte. (2015). *The top 10 issues mining companies will face in the coming year.* Retrieved from https://www2.deloitte.com/content/dam/Deloitte/global/Documents/Energy-and-Resources/gx-er-tracking-the-trends-2016.pdf

EY. (2016). *Top 10 business risks facing mining and metals, 2016–2017.* Retrieved from http://www.ey.com/gl/en/industries/mining---metals/business-risks-in-mining-and-metals

Grayson, D., & Hodges, A. (2003). *Compromisso Social e Gestão Empresarial* [Social Commitment and Business Management]. São Paulo: Publifolha.

Institute of Risk Management. (2003). *A Risk Management Standard.* Retrieved from http://www.theirm.org/publications/PUpublications.html

Kerzner, H. (2006). *Gestão de Projetos: As Melhores Práticas* [Project Management: Best Practices]. São Paulo, Brazil: Bookman Publisher.

Moura, A. A. (2013). *Gerenciamento de riscos associados a stakeholders em projetos de mineração. (Monografia de especialização)* [Management of risks related to stakeholders in mining projects. (Specialization Monograph)]. Belo Horizonte, Brazil: Pitágoras University.

Peixoto, A., Silva Neto, J. C., & Mazzuchelli, N. (2010). O Processo de Gerenciamento de Stakeholders em Projetos: Desvios da Prática Brasileira [The Process of Management of Stakeholders in Projects: Deviations from Brazilian Practice]. In *Anais, 5th Project Management Congress and 2nd Project Management Seminar in the 3rd Sector.* Belo Horizonte, Brazil: Academic Press.

Project Management Institute. (2013). *Project Management Body of Knowledge (PMBOK)* (5th ed.). Project Management Institute.

Silva Neto, J. C. (2011). *Avaliação de maturidade no gerenciamento de projetos em uma empresa de mineração em Minas Gerais* [Evaluation of maturity in the project management in a mining company in Minas Gerais State] (Masters dissertation). Fundação Mineira de Educação e Cultura, Graduate Program in Administration, Belo Horizonte, Brazil.

# Chapter 9
# How to Apply the Model to Measuring Complex Engineering Projects


## ABSTRACT

*In this chapter, the authors present the proposal of the questions and the final model which will be applied in the company in order to obtain at the end the result of the maturity level. These questions, as explained in the Chapter 2, came from a mix of project management methods and good practices and from the nature of the pillars of complex projects. The chapter also introduces the questions, explains the pillars in which the questions are split one by one, and finally presents the questions. This is one of the most important chapters in the book because it introduces the proposal to measure maturity in complex projects. The reader can easily perceive by looking over the questions and pillars that this is a different model proposed, which now includes some different pillars such as stakeholders, environment, engineering, and others. The main idea of this chapter is to introduce and explain to the reader the origin of the questions which will be, at the end, the model proposed for the book.*


DOI: 10.4018/978-1-5225-5864-4.ch009

## MATURITY IN PROJECTS

Maturity in project management is directly related to an organization's ability to manage its projects. Companies that practice project management are constantly evolving and maturing, with an intuitive relationship between maturity and success (prado, 2008).

According to Prado (2008), the occurrence of a greater maturity in the management of projects of an organization provides more predictable results. A project maturity evaluation model consists of a mechanism that has the possibility of numerically quantifying the company's ability to manage its projects.

Among the most used maturity models in the world, we can quote the following:

- OPM3, which is an abbreviation of the "Organizational Project Management Maturity Model", which is a maturity model from the PMI - Project Management Institute and which makes it possible to evaluate the presence of best practices within the organization. OPM3 consists of a set of best practices, which are required to be fulfilled for the improvement of a Domain and Internship. The model is also based on 3 elements:
    - **Knowledge:** Based on PMI's OPM3 guide, it has Best Practice components with their corresponding Capacity, Output, and Performance Indicator Items;
    - **Evaluation:** The model has a computerized tool for evaluation, comparing a standard and defining the current stage of maturity of the company;
    - **Development:** Consists of executing a plan of action through the items identified in the Evaluation element.
- The CMM - Capability Maturity Model is a widely used IT project focused maturity model, one of the first maturity models;
- The PMMM Project Management Maturity Model is a model created by Dr. Harold Kerzner, which is fairly reliable as it has been tested in various industries around the world. This model aims to define the current stage, to plan and implement actions for the gradual development in project management. Practically focuses on change management and the process of shifting the organization's culture to the implementation of project management practices;

- The Prado-MMGP Model (Maturity Model in Project Management) is presented as the only Brazilian contribution to the maturity evaluation, because it was prepared by a Brazilian, the consultant Darci Prado. This model originated through a survey conducted between October and December 2005 through the website www.maturityresearch.com and answered by 261 professionals of all types of Brazilian organizations. It is composed of a questionnaire with forty questions where it is sought to relate the maturity of the organization with its ability to execute projects successfully.

According to Prado (2008), the average and current maturity of Brazilian industries is in the range of 2 to 3, on a scale of 1 to 5, with 1 being where there is no practice and 5 with the maximum degree, that is, existence of all practices:

*It is expected that a Project Management maturity model will be able to assist in the establishment of an integrated growth plan of the organization within the context of Project Management. (Prado, 2008: 45)*

Kerzner (2003, p. 53) points out that "Project Management maturity is the development of systems and processes that are inherently repetitive and guarantee a high probability that each of them is a success".

The maturity model also represents how project management training should evolve over time. For this, the transfer of knowledge from project to project, and to the organization as a whole becomes fundamental. However, although the maturity models presented are of great importance in the project management market, they do not introduce important pillars in complex projects such as: environmental management process, geology, relationship with communities, etc. In previous chapters we talked about stakeholders. Before presenting the proposed model itself, we will go through some sectors that are new in maturity models in order to contextualize the model.

## ENVIRONMENTAL MANAGEMENT IN COMPLEX PROJECTS

Nowadays, it is perceived that the term "environment" has caused changes in large corporations. This is due to several factors, whether be they by environmental norms, be they by perceptions that their products are passing to consumers. It ends, then, by bumping into factors as a competitive advantage, as differential in relation to competitors. Among the main ones, sustainable marketing is mentioned here as a differential. But, how to seek this marketing together with the market? In the case of mining companies, the solution to minimize the impact caused by their operations will be the use of environmentally correct project management.

Over time, as an example of a complex project source, mining companies were not pressured by the "damage" done visually during their mineral extraction process. The process of mineral extraction, better known by the term "mining", has, over the last few years, been treated as central to the growth of the country. The term mining is understood to be associated with any type of extracted ore and its subsequent process chains, such as: iron ore (steel, etc.), limestone (cement, etc.) and, bauxite), etc.

The dependence on energy sources for mine supply, environmental devastation, waste of water resources, pollution of rivers, oceans, seas, land and air, as well as other innumerable actions that affect the environment have caused great impacts to modern society.

Negative impacts such as those mentioned above that have gained such a remarkable proportion that reached a point where large groups such as Green Peace and SOS Mata Atlântica have increasingly been able to gain power through the awareness of society related to these negative impacts and, still rather, to contribute to the irresponsible actions of some mining companies to be known and pressured by society.

Thus, from these pressures imposed by society, the execution of complex projects that require the environmental licensing process ends up putting pressure on companies to seek ecologically sustainable alternatives to operation, which, on the other hand, are presented as high-cost alternatives for implementation, and also a certain barrier to its realization.

However, competition for the production of cheap ore among the various companies participating in the sector presents the environmental management, also known as sustainable development, as a primordial tool and a differential for the construction of its strategy in the market of action.

Although the term "marketing" is widespread today, it is still interpreted as a deceptive action to the consumer even though it is through perceptions that products launched from companies that require complex projects with necessary environmental licensing, do not always fulfill their role within the consumer market. In this context, marketing studies and theories preach quite the opposite.

Often, this type of misinterpretation is caused by misinterpretation between the concepts of "selling" and "marketing." During the sales process, these companies try to guide the consumer to the specific examination of the final product and, thus, try to convince him to make the purchase.

Like Engel et al. (2000) mentions, the purchase process is guided by several steps to the purchase itself, passing through events such as: recognition of the need, internal and external analysis in the search of information, analysis of the differentials until the purchase materialization. This attempt involves typical marketing decisions, such as aggressive price policy, sales promotion, value-added product, sales force training, and advertising, among others.

According to Levitt (s / d), in order to keep a client in his or her wallet and preserve it, it is important the fact that the company convinced about the importance of this final product, consequently causing its satisfaction. In order to make this a reality, it is essential that the company's employees understand that their jobs are only possible to serve the final consumer and that, for their specific engineering area, it is always important to think about projects that support environmentally friendly equipment and operations, in order to, in the future, to have this specific item as a competitive advantage.

Considering this scenario, Kotler (2000, p. 47) proposes "social marketing", which determines that the organization's role is to identify "needs, desires and interests of the final consumer", seeking to deliver "guaranteed satisfaction compared to competition, in order to preserve it and improve its well-being and society. "When addressing social marketing, there is a discussion about the incorporation of alternatives to social problems in marketing strategies, in order to increase the satisfaction and final acceptance of mining products. It is through socially responsible marketing that marketing aimed at sustainable development is revealed:

*Sustainable marketing is: a way of designing and executing the exchange relationship, in order to be satisfactory for stakeholders, society and the environment, through the development, valuation, distribution and promotion of goods, services or ideas - by one party to the other party who needs them - in a way which, by helping to preserve and improve the environment, contributes to the sustainable development of the economy and society. (Calomarde, 2000: 22)*

Thus, environmentally correct companies seek through sustainable marketing what we call a "green company", understanding with this concept that the company is focused on sustainable growth and the preservation of the environment in the supply of its final products, logically, in respect to the society where it operates.

In this way, a proposal will be presented, which will refer to the importance that good project management, initially oriented towards environmentally correct alternatives, eliminates the onerous costs caused by the attempt to absorb and maintain customers by marketing sustainable development. This also eliminates the high investments of the many remodeling to fit the "green" profile and so many costs with marketing adequacy. In addition, the big gain will be that the company will already be born in the market with a vision of "green industry", not having great future obstacles related to the need to reposition its image.

## INTRODUCTION OF THE MODEL FOR MEASURING MATURITY IN COMPLEX PROJECTS

The proposal for the model is that it is descriptive, with a quantitative and qualitative nature, through studies that will be presented later in other chapters of the book. It refers to the qualitative phase of the analysis of the methods and practices adopted in the case studies and, to quantitative, the descriptive statistical treatment of the information obtained through a semi-structured questionnaire that composes the proposed model.

According to Triviños (1987), descriptive studies are used to describe in detail facts or phenomena, as well as to be used to establish associations between variables. For Malhotra (2006), the descriptive studies assume that the researcher already has some idea about the research problem and, therefore, is based on the detailing of specific hypotheses and the specification of the necessary information.

### The Maturity Model

The proposed model encompasses nine pillars: PMO (14 questions), Project Management (34 questions), People Management (9 questions), Methods and Practices (8 questions), Structure (6 questions), Performance and Results (6 questions), Engineering and Geology (25 questions), Environmental Management (9 questions) and Stakeholder Management (9 questions), which can be seen at the end of this chapter.

The questionnaire consists of closed and open questions. There are a total of seven levels of maturity ranging from level 0 - None to the maximum level of maturity that corresponds to level 6 - Strategic. Thus, the scoring scale for response of each question of maturity research is composed of 0 - Disagree and 7 - Agree. The suggestion is that the application and the availability take place by sending and returning by email or through electronic questionnaires available on various websites. Responses are individual and collected shortly after completion. The responses of the sample participants were recorded in the form of an Excel spreadsheet and compiled into a final result format.

In order to guarantee the quality of the selected questions and their calibration in the market, a case was chosen that consists of a large mining company, with operations located both in Brazil and abroad. The results and discussion of this case will be presented in the chapter on the mining sector.

## How Model Data Should Be Treated

The instrument of data collection to be used, as suggested, should be by sending the questionnaire by e-mail to the selected participants in the company. To this end, the personal data of the interviewees must be protected, as well as the names of the companies in which they execute projects, unless the application of the same is characterized by an internal application.

The data collected by the quantitative research should be treated by means of a simple arithmetic mean obtained for each of the nine independent pillars. At the end, a score between 0 and 7 is obtained, presenting the individual maturity for each separate pillar, the maturity of the company being the simple arithmetic mean of the nine pillars.

To exemplify the calculation we have: assuming that one of the pillars has 10 evaluation questions and that 4 were answered with the evaluation 0 - I disagree and 6 questions answered with the evaluation 7 - I agree. Thus, we will have as the maximum score of this pillar the final grade of: $4 \times 0 + 6 \times 7 = 42$ points. As in this pillar, the number of questions is 10, we will have as pillar maturity the evaluation of: $42/10 = 4.2$.

Ao obter o resultado final, deve-se ir na tabela abaixo e procurar pela localização da nota associada. Além da maturidade final da empresa, é importante entender os níveis de maturidade individuais para cada pilar. Isso possibilita ao gestor entender os pilares fortes assim como os pilares que necessitam desenvolvimento na empresa.

When obtaining the final result, one must go in the table below and look for the location of the associated note. In addition to the company's final maturity, it is important to understand the individual maturity levels for each pillar. This enables the manager to understand the strong pillars as well as the pillars that need development in the company.

As a measure of the company's final maturity, the calculation should be done by summing the maximum score of each pillar and, at the end, dividing this value by the number of pillars that are 9. For this, it is suggested the use of An Excel worksheet.

Regarding the treatment of the data obtained in the application of the model, through the open questions, the data should be grouped according to the perception of each interviewee in relation to the project management process of the company. Open questions are unstructured questions, to which the respondent must respond in his own words.

According to Malhotra (2006), the use of multiple choice questions presents main advantages the ease of response and tabulation. On the other hand, it presents as disadvantages the possibility of bias and the difficulty of developing effective options.

## Maturity Measurement Scale in Complex Projects

Some maturity models focus specifically on the capacity of project management as a whole, not going into specific details and themes such as: engineering, environmental licensing, stakeholder mapping - which are characteristic complex enterprise themes.

*Table 1. Final score by maturity level*

| MATURITY | RESULT |
|---|---|
| **Unexistent** | 0>X>1 |
| **Conceptual** | 1>X>2 |
| **Programmed** | 2>X>3 |
| **Standardized** | 3>X>4 |
| **Controlled** | 4>X>5 |
| **Implemented** | 5>X>6 |
| **Strategic** | 6>X>7 |

Source: Author's Proposal

Thus, our intention is to propose a maturity model that, besides the project management processes, also encompasses critical success factors such as: environmental management, stakeholder management, engineering and geology management, etc., giving rise to a way of assessing maturity in managing complex projects.

In this proposal of maturity model in complex projects, it focuses on six possible levels for classification of the company according to its evolution and adoption of the practices in project management of this specificity. In Figure 1, these six levels of the proposed scale are presented.

As can be seen in the picture, the scale follows an inverted form: from top to bottom, as the company evolves its maturity in the management of its projects, it goes from a conceptual level to a level where project management is part of its strategic planning.

Each level represented in the pyramid above translates a value for the company when evaluated, as follows in detail:

- **LEVEL 1 - CONCEPTUAL:** The company knows what project management is; has some initiatives in this area; some tools like management software already exist; the application of project management is carried out by professionals in isolation; few employees know the standards of project management; there is no structured methodology;

*Figure 1. Scale of the level of maturity in mining projects*
*Source: Author's Proposal*

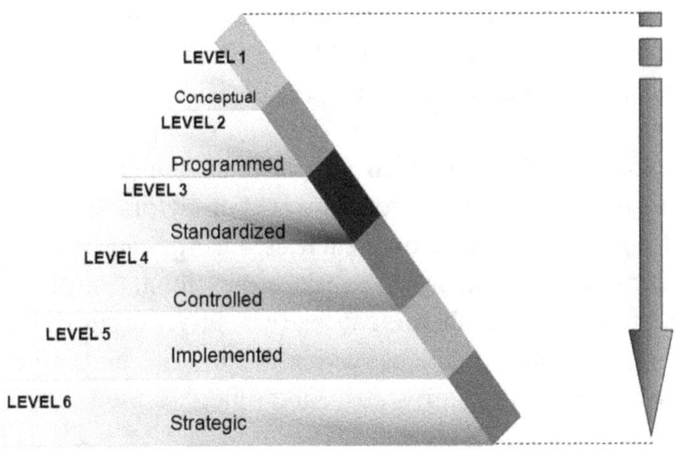

- **LEVEL 2 - SCHEDULED:** The company's top management recognizes the need for the project management standard; recognizes that this should be implemented in the short term; seeks qualified human resources and also seeks to disseminate the culture of project management within the company, as well as, seeks the standardization of project management processes;
- **LEVEL 3 - STANDARDIZED:** The company is already promoting standardization of a methodology in project management, including the process of training hierarchical levels. At this level the company begins the process of restructuring and adaptation of the structure to attend the project management; We start thinking about the implementation of a PMO - Office of Projects, as well as in management software;
- **LEVEL 4 - CONTROLLED:** The top management of the company is already committed to the need for a standardized project management methodology; all processes are implemented as follows: change management, lessons learned, assessment and management by indicators, trained managers ahead of mining projects. In short, there is already a well-disseminated culture;
- **LEVEL 5 - IMPLEMENTED:** The company already has a methodology implemented and in use by all levels of the organization; all the projects and programs of the company are managed by this methodology; already exists a PMO implanted, supporting the corporate management of the projects; people are already trained and well-versed in the methodology and process of project management;
- **LEVEL 6 - STRATEGIC:** The company's top management aligns all its strategic objectives with the projects being developed; the PMO is the center of standards and methodology in project management, as well as, there is an assessment of the maturity of the company; all projects have well-defined and PMO-controlled levels of management; indicators are generated to support decisions.

The evolution of maturity in project management of a company is not something easy. This depends on a myriad of factors that, for the most part, involves higher scale decisions, such as: changes in structure, changes in standards of conduct, hiring of expert professionals, suitability of existing facilities, etc. Often, such factors can lead executives to question the application of some management practices, which in turn, are under pressure to make their investments with quality, costs and time:

*Entrepreneurs and executives from the most diverse types of organizations have awakened to the advantages of investing in the development of an enabling and facilitating organizational context so that their projects can result in predictable returns, thus enhancing the organization's strategic management. In this sense, a myriad of new concepts, products and services have dominated and seduced managers who find themselves faced with the problem of discerning and separating what is really useful from what may not be, during the process of creating a favorable environment to the proper execution of projects and organizational initiatives. (Soler, 2008, p.1)*

In order to inform the difference in levels and evaluate the evolution of the company, we created what is called a cross-sheet as shown below, using the following factors: Information, Specialization, Application of Methods and Standardization. For each of the levels presented, a qualitative score is assigned, which differs from the other levels, according to its position in the maturity pyramid of the proposed model.

It is observed in the table that, as the company evolves from a level of conceptual maturity to a strategic level, the questions Information, Specialization, Application of Methods and Standardization are becoming implanted and as part of the strategic planning of the organization.

## PILLARS OF MATURITY IN COMPLEX PROJECTS

In the previous section, we presented a proposal of maturity levels that guide the maturity of organizations. In this section we propose to present the main pillars used to judge the classification of the company's maturity.

*Table 2. Classification of maturity levels*

| LEVEL | INFORMATION | SKILLS | METHODS | STANDARDS |
|---|---|---|---|---|
| **Conceptual** | Initial | Unavailable | Unavailable | Unavailable |
| **Programmed** | Complementary | Initial | Isolated | Isolated |
| **Standardized** | Complementary | Initial | Some Methods | Some Standards |
| **Controlled** | Applicable | Existing | Methods available | Standards available |
| **Implemented** | Applicable | Existing | Consolidated | Consolidated |
| **Strategic** | Applicable | Existing | Supported by the Board | Supported by the Board |

Source: Author's Proposal

These pillars were developed within a judgment of industry specificity during the management of complex projects. They are crucial and common to any and every project considered complex, as well as being critical during the management and implementation of the enterprise:

*Observations and reports from real cases show that countless projects ... are perceived as 'failures', although they have been completed on time and budget and have met all the technical specifications previously established. In the same way, other projects are considered 'successes', even though they have failed to comply with such items. (Robic & Sbragia, 1995, page 45)*

It should be noted that all the pillars generally involve the management of a mining enterprise, as explained below:

- **Project Management:** Basically based on the nine areas of knowledge of the PMBOK (Integration, Scope, Cost, Time, Quality, Human Resources, Communication, Risks and Supplies). It focuses on the identification and recognition of the company of these values;
- **Performance and Results:** In this pillar, the main tools that lead the project and generate its results are presented;
- **Methods and Practices:** Focuses on the existence of a methodology in project management, i.e., process mapping, activity flow;
- **Engineering and Geology:** In this pillar, the existence and evolution of the engineering levels is observed, as well as the probing campaigns to know the material in which the project will be inserted. This is one of the most important pillars to support the maturity of the company. In it lies all the intelligence that will support all other pillars. As an example, it provides information for environmental licensing processes, stakeholder mapping, etc.;
- **Organizational Structure:** Seeks to identify if the structure, which will lead the project, is the best or if adjustments are needed to form an organization for better management;
- **People Management:** This pillar focuses on the career development of the people involved in the project;
- **Project Management Office:** This pillar aims to identify the existence and recognition of the need for a project office capable of supporting the entire methodology for project management;

- **Stakeholder Management:** It is sought to conduct the identification and management of the main stakeholders of the project, whether they are positively or negatively impacted;
- **Environment Management:** This pillar represents the critical path of most mining projects. In this part, the best practices for reaching the main licenses are discussed: prior, installation and operation for the project. It also seeks to have as practice the development of the whole project based on sustainability.

## STRUCTURED QUESTIONNAIRE FOR MATURITY ASSESSMENT

Based on the previously defined pillars, for each of them, we applied a structured questionnaire to managers or project managers in order to collect their perception about the level of maturity of the projects or organization. The scale to be used for completing the answers constitutes 0 if you disagree with the question or 7 if you agree with the question. Only responses are allowed or 0 or 7. Any number typed other than this will not be accepted

*Figure 2. Pillars to support maturity levels*
*Source: Author's Proposal*

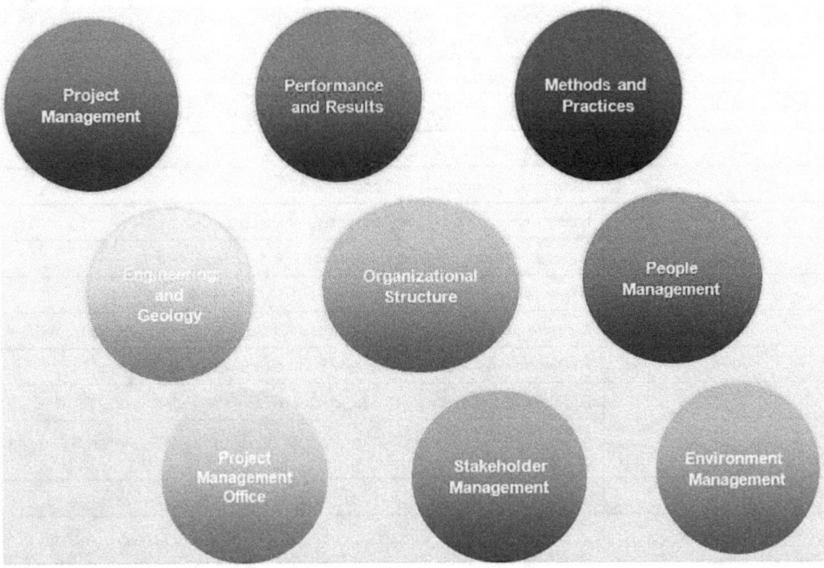

*Table 3. Project management*

| Id. | Question |
|---|---|
| 01 | Is there any document to formalize the existence of the project in your area? |
| 02 | Does the project existence document adequately describe the project output? |
| 03 | Is there a document that can register the assumptions (prerequisites necessary for you to carry out the work) and constraints (dates imposed by the client, limitation of funding, limitation of resources) of the project? |
| 04 | Is there a Project Analytical Framework (EAP) or WBS? |
| 05 | Is there any tool for formalizing the client's acceptance? |
| 06 | Is there a baseline against which project progress, costs and deadlines are measured? |
| 07 | Is there an established process for project change control? |
| 08 | Is there a process of identifying predecessors and successors of project activities? |
| 09 | Is there a separation of project activities according to each schedule of the resource to be used (business days, holidays, etc.)? |
| 10 | Is there a process for identifying and prioritizing the critical path of the project? |
| 11 | Is there a baseline recording of the project at the end of the schedule development? |
| 12 | In developing the schedule, do all activities have resources? |
| 13 | Is there a formalization process of generating provisions (contingencies) for project costs? |
| 14 | Is the process of formalizing a matrix of responsibilities for the project? |
| 15 | Is there a formalized process of generating and transmitting project information during its execution (Communications Plan)? |
| 16 | Is there a mapping that details who needs the information, when it will be needed, how it should be defined, and by whom? |
| 17 | Is there a place with phones, addresses, e-mails and contacts from the entire team involved? |
| 18 | Is there a formalized process for defining the schedule of project meetings? |
| 19 | Are meetings pre-scheduled from the start and on the agenda of the entire team? |
| 20 | Does each team member know what information to bring to the follow-up meeting? |
| 21 | Is there a formal project budgeting process by grouping estimates of activities, resources, hiring, contingencies, etc.? |
| 22 | Is there a formalized risk identification process for the project? |
| 23 | Is there a qualitative and quantitative project risk assessment process? |
| 24 | Is there a project risk prioritization process? |
| 25 | Is there a risk planning process for the project? |
| 26 | Is there a process to identify who is responsible for the project risk? |
| 27 | Is there a risk management and monitoring tool for the project? |
| 28 | Is there a process for identifying project product requirements? |
| 29 | Are the project product requirements translated into the work required to develop this product? |
| 30 | Is there a document that can formalize the work of the project (Scope Statement)? |
| 31 | Is there a formalized audit plan for quality requirements for the project (Quality Plan)? |
| 32 | Is there formalized planning for the provision and contracting of project resources (engineering, equipment, etc.)? |
| 33 | Is there a formalized contracting cycle for large project deliverables (Bid Cycle)? |

Source: Author's Proposal

*Table 4. Performance and results*

| Id. | Question |
|---|---|
| 01 | Are deadlines met as planned? |
| 02 | Is the quality of the delivered projects according to the customer's requirements? |
| 03 | Are project costs met as planned? |
| 04 | Is there a monitoring of customer satisfaction with the project product? |
| 05 | Is there a process for evaluating the performance of the project manager in the conduct of the project? |
| 06 | Is there a process for evaluating the project team? |

Source: Author's Proposal

*Table 5. Methods and practices*

| Id. | Question |
|---|---|
| 01 | Before the start of the project is a feasibility study for the investment carried out? |
| 02 | Is there a practice of benchmarking projects in relation to projects developed by similar companies? |
| 03 | Is there practice of contracting consultancies to identify and evaluate the project's maturity for successful projects (IPA - Independent Project Analysis, etc.)? |
| 04 | Is there practice of hiring consultancies in the evaluation and optimization of budget? |
| 05 | Is there a formal process of closing contracts open during the project? |
| 06 | Is there a formal process of collecting the lessons learned during and at the end of the project? |
| 07 | Is there a formal process of cost control through tools like VAE - Value Added Estimation? |
| 08 | Is there a formal operational risk assessment and survey process (HAZOP - Hazard and Operability Studies)? |

Source: Author's Proposal

by the spreadsheet. For this, each of the questionnaires, divided among the previous 9 pillars, is shown in Table 3.

## CONCLUSION

The application of the proposed model for measurement of maturity in complex projects is an instrument of simple manipulation and application. As a suggestion in your application, it is the tip of some important steps that can be noticed in the section where we will show the case studies proposed in the book. Some important steps:

*Table 6. Engineering and geology*

| Id. | Question |
|---|---|
| 01 | Is there a formalized process that every project developed in the organization should go through the phases: conceptual, basic and detailed? |
| 02 | Is every project being run by the organization going through the conceptual, basic, and detailed engineering phases? |
| 03 | Is there a formalized process for the drilling and auditing of mineral resources for the execution of engineering? |
| 04 | Is there a formal process that all conceptual engineering must begin with the first results of exploratory surveys? |
| 05 | Is there a formalization and submission of the conceptual engineering information to the Environmental area for environmental licensing? |
| 06 | Is all Conceptual Engineering developed based on a mapping of biotopes (areas impacted by the environment) and surface? |
| 07 | Are all process tests (Technological Development) formally sent to Engineering to start Basic Engineering? |
| 08 | Is there a formalized process for reviewing engineering quality (Peer Review) before moving from one phase to another (e.g., Conceptual Engineering for Basic Engineering)? |
| 09 | Is there a formalized engineering development culture in 3D technology? |
| 10 | Is there a process of evaluation, prior to the beginning of the Detailed Engineering, on the detailing of the project through the modularization of the construction? (Construction in Modules) |
| 11 | Is there the process of generating a Master Plan for the project after Conceptual Engineering? |
| 12 | If yes for answer 11, is there a formalized process of freezing this master plan and communicating it to all project stakeholders? |
| 13 | Is there a formalized process of generation of Descriptive Memorials, Quantities Worksheets, Datasheet and Technical Specifications for the Supplies area? |
| 14 | Is there a formalized process of participation of the Supply area during the evolution of the Basic Engineering for Detailed Engineering to determine the type of contracting of the Supply, Assembly and Implementation of the project? |
| 15 | Is there a formal process of realizing and studying tradeoffs for the lowest cost for utilities (water, gas, electricity, logistics, etc.)? |
| 16 | Is there a formal process of realization and study of opportunities (tradeoffs) regarding the location of the project (surface workers, communities, logistics, etc.)? |
| 17 | Is there a formal process for evaluating and validating each engineering phase (Conceptual, Basic, and Detailed) before moving from one to the other (Front End Loading Portions of Independent Project Analysis)? |
| 18 | At the end of the Project's Conceptual Engineering, is there a formal process for generating a Prefeasibility Study? |
| 19 | At the end of the Basic Engineering of the Project, is there a formal process to generate a Feasibility Study? |
| 20 | During the execution of the Conceptual Engineering, is there a formal process of identification and first contacts with impacted landowners as well as local communities? |
| 21 | During the execution of the Conceptual Engineering, is there a formal topographic survey of the region as well as the generation of orthophotos for the project's master plan? |
| 22 | After Conceptual Engineering, is there a formal process of generating investment estimates at an accuracy level of -15% to + 25%? |
| 23 | After the Basic Engineering, is there a formal process of generating investment estimates at an accuracy level of -10% to + 10%? |
| 24 | Is there a formal closure plan process? |
| 25 | Is there a formal process for identifying VIPs - Value improving practices? |

Source: Author's Proposal

*Table 7. Organizational structure*

| Id. | Question |
|---|---|
| 01 | When a project starts, is there a project manager assigned by the organization to lead the project? |
| 02 | Is there an area focused on project planning? |
| 03 | Is there a formalized position of PROJECT MANAGER within the organization? |
| 04 | Is the Project Manager's formalization of power for similar functional positions (HR Manager, Operations Manager, etc.)? |
| 05 | Is there formalization of areas within Project Management such as: Planning, PMO, Cost Control, etc.? |
| 06 | Are all internal areas of the company formally communicated from the beginning of the project and are their representatives formally assigned to the project? |

Source: Author's Proposal

*Table 8. People management*

| Id. | Question |
|---|---|
| 01 | Is there a formal career development process for the project manager and team? |
| 02 | Do follow-up meetings have a schedule and a fixed format? |
| 03 | Does each project member receive their work with instructions and in advance? |
| 04 | Is there a formal process of variable compensation for project team performance? |
| 05 | Is there a formal process for evaluating project team performance? |
| 06 | Is there a formal organizational chart of the project team, delegations, etc.? |
| 07 | Are all the human resources needed for the project formally sent to the HR area to arrange recruitment? |
| 08 | Are all interpersonal skills of the project team managed during the project? |
| 09 | Are all conflicts between the project team formally addressed? |

Source: Author's Proposal

1. Always elect an audience that knows about the existing methodology of project management as it has already applied it in past projects or projects in progress. This will help in the correct calibration of the target public of the research, so that the results are the most accurate;
2. When defining the target audience, promote a group meeting in order to explain the reasons for the maturity measurement, how the model is divided, etc. If possible, explain the question, leaving the scoring rules clear, etc.;

*Table 9. Project management office*

| Id. | Question |
|---|---|
| 01 | Is there a flow for the arrival in your area of information about the existence of a project or a contract? |
| 02 | Is there a dedication of the professionals to the management of the projects? |
| 03 | Is there a tool for estimating effort (time) and project costs? |
| 04 | Is there any tool for estimating project resources? |
| 05 | Is there any tool for formalizing the client's acceptance? |
| 06 | Is there an established process for formal acceptance of each deliverable of the project? |
| 07 | For the development of project documentation, is there a repository of past project templates for reuse in new projects? |
| 08 | Is there an area focused on the physical and financial control of the project? |
| 09 | Is there a visibility room, a mural or even an intranet site where everyone can access projects in a simple way? |
| 10 | In the case of geographically dispersed teams, are the procedures / technologies used to integrate them adequate? |
| 11 | Are there managerial reports with a predefined format agreed with the client? |
| 12 | Is there a formalized process for closing a project / contract? |
| 13 | Is there a formal process for the provision of information in the closing phase (completion of work)? |
| 14 | Is there an area focused on the control and support of the project (PMO or Office of Projects)? |

Source: Author's Proposal

*Table 10. Stakeholder management*

| Id. | Question |
|---|---|
| 01 | Is there a formalized process of mapping and managing the company's internal stakeholders? |
| 02 | Is there a formalized process for managing stakeholder expectations of the project during its execution? |
| 03 | Are formal interviews conducted to identify project stakeholders? |
| 04 | Are all potential impacts caused by each stakeholder formally assessed and managed during and after the project? |
| 05 | Is there a formal process of ranking and prioritizing the degree of influence of each stakeholder on the project? |
| 06 | Is there a formalized process of classifying stakeholder groups? |
| 07 | Is there a formal process of simulating the reactions of each stakeholder to the various situations that may occur during and after the project? |
| 08 | Is there a formalized process of collecting and storing the information of each stakeholder such as: name, address, telephone, etc.? |
| 09 | Is there a formalized Communication Plan capable of identifying what kind of information, when, how and for which stakeholder to send it to? |

Source: Author's Proposal

*Table 11. Environment management*

| Id. | Question |
|---|---|
| 01 | Is there a formalized process of beginning environmental studies with the beginning of Conceptual Engineering? |
| 02 | Is there a formalized mapping process of biotopes (areas of environmental impact) to aid in the development of Conceptual Engineering? |
| 03 | Is there a formalization process of mapping the surface properties impacted by the project during the Environmental Licensing and Conceptual Engineering? |
| 04 | Is there a formal process of generating documentation for EIA - Environmental Impact Studies and RIMA - Engineering Environmental Impact Report? |
| 05 | Is there a formal process that synchronizes the development of Engineering and Environment activities? (Generation of Documentation for Environmental Licensing)? |
| 06 | Is there a formalized area for institutional relations with government agencies (IBAMA, SUPRAM, IGAM, etc.)? |
| 07 | Is there a formalized area for relationship with communities impacted by the project? |
| 08 | Is there a formalized integration of the internal areas of the company: Engineering, Environment, Institutional Relationship and Social Responsibility? |
| 09 | Is there a formal process to monitor the release of Pre-Installation, Installation and Operation licenses from government agencies? |

Source: Author's Proposal

3. Always apply the questionnaire in order to give the public time to get used to it. Usually, after the introductory meeting mentioned in point 2 above, the questionnaire is distributed immediately and 2 days is a feasible deadline.

With the data in hand, the results are easily compiled and can be structured in the best way to present the results to the high level of the company. Based on the points of weakness presented in the result, it is possible to trace actions and next steps, as well as to define next level goal, etc. This is important because in some cases investments are necessary and need to be justified at the highest level of the company.

## REFERENCES

Calomarde, J. V. (2000). *Ecologic Marketing*. Madrid: Pirâmide, ESIC.

Engel, J. F., Blackwell, R. D., & Miniard, P. W. (2000). *The Consumer Behavior* (8th ed.). Rio de Janeiro: LTC.

Kerzner, H. (2003). *Project Management: a system approach to planning, scheduling, and controlling. 8*. John Wiley & Sons.

Kerzner, H. (2005). *Using the Project Management Maturity Model – Strategy Planning for Project Management*. John Wiley & Sons.

Kerzner, H. (2006). *Project Management: The Best Practices. 2*. Porto Alegre: Bookman.

Kotler, P. (2000). *Marketing Administration the edition of the new milenion*. São Paulo: Prentice-Hall.

Levitt, T. (2000). *Marketing for Business Development*. São Paulo: Cultrix, s.d.

Malhotra, N. (2006). *Marketing Research: a planned orientation* (4th ed.). Porto Alegre: Bookman. doi:10.1108/S1548-6435(2006)2

PMBOK - Project Management Body of Knowledge. (2009). *PMI - Project Management Institute. 4*. PMI.

Prado, D. (2004). *Project Planning and Control*. Belo Horizonte: INDG TECs.

Prado, D. (2008). *Maturity in Project Management*. Belo Horizonte: INDG TECs.

Research And Metrics Measuring Capital Project (2004). *IPA – Independent Project Analysis*. EUA.

Robic, R. A., & Sbragia, R. (1995). *Success in Projects for Informatization: Evaluation Criteria and Success Factors*. São Paulo: Economia Empresa.

Soler, A. M. (2002). *Organizational Maturity and the Evaluation Model of PMI-OPM3. 4*. Rio de Janeiro: Campus.

Triviños, A. N. S. (1987). Introduction to Social Science Research: the qualitative research in education. São Paulo: Academic Press.

# Section 4

# Case Studies: Maturity, Complex Projects, Engineering, and Construction

# Chapter 10
# Project Maturity Analysis in the Mining Sector


## ABSTRACT

*In this chapter, the authors present a case study with the objective to evaluate the results of the same in the mining sector, one of the sectors in the economy with more relevant complex projects around the world and in which one can easily apply all of the pillars as discussed during the presentation of the model. To better give the reader an understanding about the idea, the authors first introduce the sector and then perform and present the results of the investigation over this case. In order to calibrate the proposed model, the authors decided to apply it by checking at the end if the result achieved was closer than what they imagined for the company.*


## PROBLEMATIZATION

Mining is one of the basic sectors of the country's economy, contributing decisively to the well-being and improvement of the quality of life of present and future generations, being fundamental for the development of an equitable society, provided it is operated with social responsibility, and the precepts of sustainable development are always present (Wagner et al., 2002). As the mining sector is characterized by investment and these investments are materialized in several projects in the industry, it can be expected that the control and evaluation of the projects reflect the profitability and profitability of the sector.

DOI: 10.4018/978-1-5225-5864-4.ch010

In view of this investment scenario, the mining sector should focus its efforts on finding technologies, procedures, methods or practices that will enable the management of its projects with a view to achieving goals and profitability. Searching for methods and practices of project management may reflect that organizations seek to learn, to know and to disseminate information, which can be understood as a model of maturity.

The proposed case study refers to a company located in Brazil, more specifically this company is located in the state of Minas Gerais. Brazil produces about 70 substances, of which 21 are from the metallic mineral groups, 45 from the non-metallic ones and four from the energetic ones. In terms of market share, the position of niobium (92%), iron ore (20%, second largest producer in the world), tantalite (22%), manganese (19%), aluminum and asbestos %), graphite (19%), magnesite (9%), kaolin (8%) and ornamental stones, talc and vermiculite, with about 5% (Barreto, 2001). According to IBRAM (2009), Minas Gerais accounts for 35% of total Brazilian production, being the largest producer in the iron ore area.

In this context, then, it is based the central question of this proposed maturity evaluation through which one intends to investigate: **What is the level of maturity in project management in an iron mining company in Minas Gerais?**

## THE MINING SECTOR

The History of Brazil is closely related to the search for and exploitation of its mineral resources, which have always contributed important inputs to the national economy, as part of the territorial occupation and national history. According to Wagner et al. (2002), the mineral sector accounted for 8.5% of GDP in 2000, that is, 50.5 billion dollars generated 500,000 direct jobs and a balance of trade of 7.7 billion dollars. an average annual growth of 8.2% in the period 1995/2000.

Brazil, according to the DNPM (2006), due to its privileged geodiversity and continental extension, occupies a prominent position due to the recognized comparative advantage of its international class deposits and mines, positioning itself competitively in the international market of mineral commodities market, with Australia, Canada, China and South Africa, competitors holding large mineral reserves.

The history of Brazil in terms of mining is extremely complex. According to Germani (2002), the first mining works were the gold mines, obtained in a mine known as Mina da Passagem, in Mariana, in 1819, by Baron de Echewege. Then, other mines were opened, the main one being the Old Mine of the Saint John Del Rey Mining Co. in Nova Lima in 1834 in Minas Gerais. The two mines mentioned above, and considering Mina Velha de Morro Velho, were, at the time, symbols of the use of technology and were considered worldwide references.

At that time, due to the lack of specialized tools, everything was considered to be very rudimentary. Most of the time, pointers and sticks were used, and in some cases even homemade gunpowder in rock drilling:

*The Mineral Sector, since it is a cross-section of the three sectors of the Primary economy (mineral exploration and mining), Secondary (Mineral Transformation: Metallurgy, Steel, Chemistry, etc.) and Tertiary (Market, Trade) has a great breadth and heterogeneity of segments, with different stages of development (pre-mining, mining, artisanal) to modern (mineral research and planned mines from the perspective of technical excellence in Geology, Mining Engineering and Mineral Economics). (DNPM, 2006, p. 23)*

Of all Brazil's mineral potential, iron ore occupies a prominent position in the country's economy. According to IBRAM (2010), it occupies the first place in the ranking of products that generate higher income through exports. Thus, it is concluded that mining in both Brazil and the world is one of the main positive factors for a country's economy.

According to Calaes (2009), in its integrated report for the MME - Ministry of Mines and Energy -, the comparative studies of competitiveness related to the chain of Geology, Mining and Mineral Transformation are not very well diffused in Brazil. "The generation of value (whether private or social) is the fundamental indicator of the effective competitive position of an enterprise, company, sector or region" (Calaes, 2009, page 7).

According to Machado and Fonseca (2010) of BNDES - National Development Bank - prices and commodity volume rose rapidly in the beginning of the XXI century, which was caused by the great demand of the growing countries in Asia, especially the China:

*Thus, particularly in the period 2004-2008, the economic activities in Brazil related to the iron ore and steel sectors, basic metals (aluminum, nickel, zinc and copper, among others), pulp and some paper segments had annual average growth of revenues well above historical averages, and a strong investment cycle was initiated, with the implementation of new units and capacity expansion of existing facilities. (Machado & Fonseca, 2010, p. 322)*

According to Ferreira (2002), according to the Ministry of Mines and Technology, Brazil ranks as one of the largest holders of iron ore reserves in the world, behind only countries like Russia, Australia, Canada and the USA:

*However, the Brazilian reserves, besides being large, have natural technological characteristics that present an advantage in relation to these other countries. In addition, the deposits are easy to drill and allow the production of large volumes at low costs. (Ferreira, 2002, p. 9)*

Regarding the capacity of Brazilian reserves, Brazil has one of the best iron ore reserves in the world:

*Brazil has large reserves of high quality iron ore. There are 17 billion tons of iron contained in 33 billion tons of total reserves, measured and indicated. Two major regions - Carajás (PA) and the Quadrilátero Ferrífero (MG) - are rich in excellent grade ore. The Carajás deposits have the ore considered the best quality in the world, due to the high iron content and the low percentages and contaminants. The quality of the ore and the infrastructure (railroad-port) provide the main Brazilian mining companies with competitive advantages, thanks to the low cost per ton sold. Brazilian competitiveness in the sector and the apparent sustainability of Chinese demand signaled a favorable scenario for the coming years, with the possibility of price increases already in 2010 and in subsequent years, as recommended by several industry analysts. (Machado & Fonseca, 2010, p. 325-326)*

According to IBRAM (2011), Brazil is the second largest iron ore producer in the world with an estimated capacity of approximately 16% of world production. The good choice of basing the proposed case on a Brazilian mining company is therefore justified.

## THE APPLICATION AND EVALUATION OF THE PROPOSED CASE

According to Triviños (1987), one of the great benefits of case study is the deepening of a certain reality. In this way, the results obtained by means of these are valid only for the specific case of the study. According to Yin (2001), the case study brings to the fore the understanding of individual, organizational, social and political aspects.

The research was framed in these questions, since it was based on the study of the maturity in the management of projects of a large iron mining company, concentrated in the state of Minas Gerais and that has one of the largest budgets in investments of expansion projects for the next five years.

### Data Collection

The evaluation was done through a field survey through a case study in an iron mining company with an office in Minas Gerais through a semistructured questionnaire containing the determinants of the constructs. The study focused on the process of managing the investment projects of this company. The unit of analysis was the board responsible for the development and implementation of the organization's projects. Already, the unit of observation, focused on sending the questionnaire that was directed to all coordinators of the projects under development, composed of approximately ten coordinators. For data collection, the field study technique was used, which consists of the application of a semi-structured questionnaire, which is the proposed model, as shown in the previous chapters.

### The Selected Case

The chosen case is a large mining company, with operations located both in Brazil and abroad. The product produced by this company is iron ore and it has one of the largest portfolios of investments announced for the next five years.

The company presents itself in rapid (frank) growth and adoption of best practices in project management. Currently, it is structuring its area of project management through the adoption of the best tools available, in order to assist in the management of its portfolio of investments.

## Treatment of Data

The instrument of data collection used was the sending of the questionnaire by e-mail to the selected participants in the company, as already mentioned above. For this, the personal data of the interviewees as well as the names of the companies in which they execute mining projects were kept.

The data collected by the quantitative research were treated using a simple arithmetic mean obtained for each of the nine independent pillars. In the end, a score between 0 and 7 was obtained, which represented the individual maturity for each separate pillar, the maturity of the company being the simple arithmetic mean of the nine pillars.

# PRESENTATION OF THE RESULTS OF THE RESEARCH

The following data will be presented in three parts: the first will characterize the profile of the interviewees related to the survey; the second, the survey and collection of research data, according to the nine proposed maturity pillars; and the third, the qualitative information collected through the questionnaire that are not related in the questions of the pillars, and some evidence obtained outside the application of the questionnaire.

## Profile of Respondents

All the interviewees are part of the project management structure of the company surveyed and were in charge of managing projects related to the company portfolio. All project managers have at least one year of experience in the company, which leads us to understand that knowledge about the management processes adopted in the company.

## Analysis of Quantitative Research Data

The research was sent to the project managers by e-mail from the company with a stipulated deadline for a seven-day return. The questionnaire was sent to ten project managers who make up one hundred percent of the company's project management structure, in order to guarantee the survey throughout the company's project structure.

Of the ten respondents, nine returned representing a contingent of ninety percent of respondents, and only one project manager did not return for the reason that they no longer belong to the company's project management structure, representing only ten percent of the respondents. The table below compiles the information collected for each project manager. The acronym PM - Project Manager - is used to anonymously identify each respondent. The table reveals the maturity perceived by each project manager surveyed for each of the nine proposed pillars. The final maturity perceived by each project manager then became the arithmetic mean of the nine pillars surveyed.

Thus, the following represents the arithmetic mean for each of the project managers surveyed. This average was calculated by summing the score of each pillar and, at the end, dividing by the number of pylons. The GENERAL column of this same table shows the final arithmetic mean of the maturity perceived by each project manager.

*Table 1. Pillar maturity evaluation*

| Item | Pillar | PROJECT MANAGER | | | | | | | | | |
|---|---|---|---|---|---|---|---|---|---|---|---|
| | | 1 | 2 | 3 | 4 | 5 | 6 | 7 | 8 | 9 | TOTAL |
| 1 | Project Management Office | 4,0 | 5,0 | 4,5 | 3,5 | 4,0 | 3,5 | 2,0 | 3,0 | 5,0 | **3,8** |
| 2 | Project Management | 2,1 | 5,4 | 3,3 | 3,9 | 2,7 | 3,5 | 2,9 | 2,1 | 2,3 | **3,1** |
| 3 | People Management | 3,1 | 3,1 | 4,7 | 4,7 | 2,3 | 6,2 | 0,8 | 2,3 | 2,3 | **3,3** |
| 4 | Methods and Practices | 3,5 | 5,3 | 6,1 | 5,3 | 3,5 | 2,6 | 2,6 | 2,6 | 1,8 | **3,7** |
| 5 | Organizational Structure | 5,8 | 1,2 | 7,0 | 5,8 | 3,5 | 2,3 | 2,3 | 4,7 | 2,3 | **3,9** |
| 6 | Performance and Results | 2,3 | 3,5 | 3,5 | 2,3 | 0,0 | 3,5 | 1,2 | 2,3 | 2,3 | **2,3** |
| 7 | Engineering and Geology | 3,9 | 3,6 | 4,5 | 3,6 | 0,8 | 3,1 | 1,7 | 2,2 | 1,7 | **2,8** |
| 8 | Environment Management | 5,4 | 4,7 | 3,1 | 3,9 | 1,6 | 5,4 | 3,9 | 5,4 | 3,9 | **4,1** |
| 9 | Stakeholder Management | 3,1 | 3,1 | 1,6 | 3,1 | 2,3 | 4,7 | 3,1 | 2,3 | 3,9 | **3,0** |
| | **MATURITY** | **3,7** | **3,9** | **4,2** | **4,0** | **2,3** | **3,9** | **2,3** | **3,0** | **2,8** | **3,3** |

Source: Author's proposal.

*Table 2. Results by project manager*

| | PROJECT MANAGER | | | | | | | | | |
|---|---|---|---|---|---|---|---|---|---|---|
| | 1 | 2 | 3 | 4 | 5 | 6 | 7 | 8 | 9 | TOTAL |
| **MATURITY** | 3,7 | 3,9 | 4,2 | 4,0 | 2,3 | 3,9 | 2,3 | 3,0 | 2,8 | **3,3** |

Source: Author's proposal

*Table 3. Final score by maturity level*

| MATURITY | RESULT |
|---|---|
| Unexistent | 0>X>1 |
| Conceptual | 1>X>2 |
| Programmed | 2>X>3 |
| Standardized | 3>X>4 |
| Controlled | 4>X>5 |
| Implemented | 5>X>6 |
| Strategic | 6>X>7 |

Source: Author's Proposal

*Table 4. Maturity scale for evaluation*

| Scale | Description | Level |
|---|---|---|
| Level 0 - Unexistent | Not developed in the company | 0 |
| Level 1 - Conceptual | The company knows what is project management, has some initiatives in this area, some tools such as management software already exist, the application of project management is carried out by professionals in isolation, few employees know the standards of project management, not there is a structured methodology. | 1 |
| Level 2 - Programmed | The company's senior management recognizes the need for the standard of project management in mining, recognizes that this must be implemented in the short term, seeks qualified human resources and also seeks to disseminate the culture of project management within the company, as well as search for standardization of project management processes. | 2 |
| Level 3 - Standardized | The company is already promoting standardization of a methodology in the management of mining projects, including the process of training hierarchical levels. At this level the company begins the process of restructuring and adaptation of the structure to attend the management of the projects, we start thinking about the implementation of a PMO - Office of Projects as well as management software. | 3 |
| Level 4 - Controlled | The top management of the company is already committed to the need for a standardized project management methodology, all processes are implemented such as: change management, lessons learned, evaluation and management by indicators, trained managers ahead of mining projects. In short, there is already a widespread culture. | 4 |
| Level 5 - Implemented | The company already has a methodology implemented and in use by all levels of the organization, all the projects and programs of the company are managed by this methodology, there is already a PMO implemented supporting the corporate management of projects, people are already trained and know well the methodology and the project management process within mining. | 5 |
| Level 6 - Strategic | The company's top management aligns all the company's strategic objectives with the projects being developed, the PMO is the center of standards and methodology in project management, as well as assesses the maturity of the company, all projects have well defined management levels and controlled by the PMO, indicators are generated to support decisions. | 6 |

Source: Author's Proposal

For the case study, the final maturity level was 3.3, which, observing the scale presented in the table below, shows that the company presents itself at a STANDARDIZED level of maturity.

With this we have, analyzing the level reached, the definitions shown in Table 4.

For the STANDARDIZED maturity level, we have according to TABLE 4, that this level represents the state enterprise as:

*The company is already promoting standardization of a methodology in the management of mining projects, including the process of training hierarchical levels. At this level the company begins the process of restructuring and adaptation of the structure to attend the management of the projects, we start thinking about the implementation of a PMO - Office of Projects as well as management software. (Proposal of the Author)*

In the chart below we can observe the maturity of each pillar, as well as the final maturity for the case study studied and its respective placement within the maturity scale.

## Qualitative Research Data Analysis

In addition to the quantitative data presented in the previous section, the research gathered qualitative data through the collection of the impressions

*Figure 1. Overall maturity rating per pillar*
*Source: Research data.*

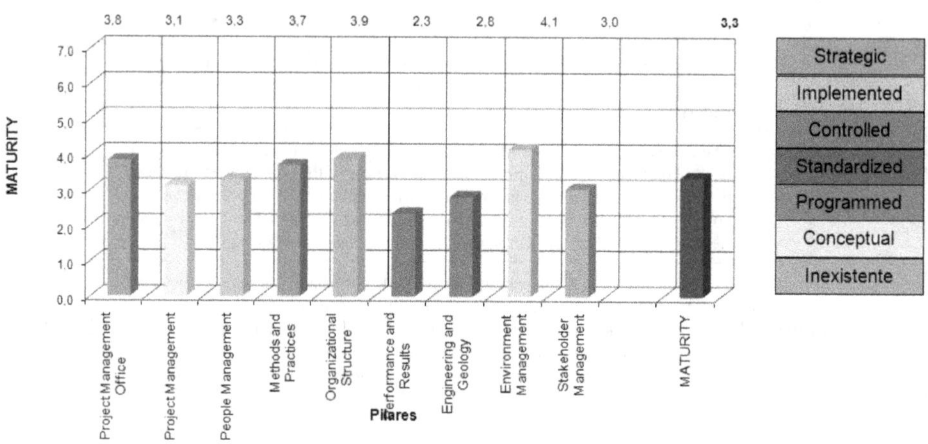

of the project managers on each observed pillar. The data will be presented for each of the nine pillars observed.

## Pillar: PMO – Project Management Office ou Escritório de Projetos

The PMO pillar - Project Management Office - had as its main objective to collect information on the existence or presence of a project office within the research company. Regarding the level of maturity. This pillar received the score of 3.8 as per survey, that is, it is currently at the STANDARDIZED level. The main observations are as follows:

- **Project Manager 3: "**Some tools began to be used as the presentation of scope for the internal and external audiences. This tool aims to provide a pre-presentation of the project to the internal team and to the proponents the execution of the Project. "
- **Project Manager 9:** I understand that in relation to items punctuated with "0", it can be said that the processes and tools are in the development and deployment phase, they are not operative bringing the expected benefits. "

In this way, it is observed, through the data collected above, that the project office already has some tools for the management of projects in use and that it is still in the phase of implementation of other tools, according to the definition of the maturity level, characterized for this item.

## Pillar: Project Management

With respect to the Project Management pillar, its objective is to identify the presence of tools that support and assist in the management of projects. In the case of this pillar, the maturity evaluated was 3.1, also in the STANDARDIZED scale. The main observations made by project managers are as follows:

- **Project Manager 3:** "Some questions started to be implemented for the projects, but the information is still restricted to the PMO."
- **Project Manager 9:** "There are development initiatives, however, they are not in place.

It is observed "from the information collected that some tools, although implemented, are not being used by the project managers because they were not well diffused by the PMO area. It is also noted that the company's project office continues to develop new tools to support the company's project management.

## Pillar: People Management

In the case of the People Management pillar, its main function is to identify the presence of tools that allow the management of the human resources of the projects. In this case, its maturity assessment was 3.3, standing on the STANDARDIZED scale. The following are the main perceptions regarding this pillar during the survey:

- **Project Manager 3:** "We are improving interpersonal interactions with more focused meetings to address internal conflicts."
- **Project Manager 9:** "There are initiatives, however, are not fully disseminated or implemented."

In this case, it is observed that the application of tools, such as meetings, is helping the team to focus on the project objectives, as well as the project office continues to develop new techniques to support project managers in the management of human resources.

## Pillar: Methods and Practices

The main objective of the Methods and Practices pillar is to identify the existence of a well-defined flow of processes to assist project managers in the execution and conduct of their projects. The maturity level evaluated for this pillar was 3.7, also at the STANDARDIZED level. The main information collected by the research is presented below:

- **Project Manager 3:** "In certain projects, some tools have been used, which in my view we still can not qualify as a process."
- **Project manager 7:** "Include the process formalized in all items, although the activity is exercised; but you do not have the location set to control the status of the Information. "

In this way, it is observed that the tools being developed by the project office and applied by project managers are still in the process of being consolidated to become part of the company's project management process.

## Pillar: Structure

The Structure pillar seeks to identify the presence of an adequate organizational structure for the management of the company's projects. In this case, the maturity level evaluated was 3.9, being the highest level of maturity per evaluated pillar, which leads us to understand that the company is in the structuring phase to fit its structure format . The following are the perceptions for this item:

- **Project Manager 3:** Some items in the structure begin to exist, but interaction with the disciplines is still premature.

It is observed the presence of a structure in the company, but the fundamental point is the initiative of interaction between the areas.

## Pillar: Performance and Results

In the case of the Performance and Results pillar, its main objective is to identify tools for monitoring the main results, as well as the progress of the main project activities. Its maturity level is 2.3, that is, at a SCHEDULED level. The following is the data collected by the research:

- **Project Manager 3:** "Pending for performance and results; still have to look for current and major projects."

This was the pillar with the lowest maturity evaluation, of the nine pillars evaluated. It is understood that some control tools that are being implemented by the project office for performance control are not being used, or do not present a good efficiency in the control.

## Pillar: Engineering and Geology

The Engineering and Geology pillar aims to identify if the organization is following the proposed process of engineering phases aligned to the life cycle for the geological investigations. His maturity rating was 2.8, SCHEDULED level. The data in this pillar are as follows:

- **Project Manager 3:** "Formal processes should be passed on to all project areas by the project manager, whether in meetings or in pre-defined documents."

It is observed that, although there is a formalized process within the company, it must be communicated and instituted within the organization.

## Pillar: Environmental Management

In the case of the Environmental Management pillar, its main focus is to identify the alignment between the engineering team and the team responsible for environmental licensing, so that the environmental management process gains speed, being provided by all the necessary information at the right time. Its maturity level is 4.1, entering the CONTROLLED scale. The following is the research data:

- **Project Manager 3:** "The interface between Engineering and Environmental Management needs to be improved. If there are processes that I disagree with, I intensify the lack of interaction between the areas. "

It is observed that this is the most mature pillar of the maturity evaluation. This is due to the fact that the company dedicates efforts to concentrate the two areas, engineering and environment under the same structure, which could be identified during the research.

## Pillar: Stakeholders Management

The main objective of the Stakeholder Management pillar is to identify the presence of tools and processes that assist the company in the identification and knowledge of all stakeholders that are affected positively or negatively by the project. For this case, the maturity stood at 3, its level being identified as STANDARDIZED. The following are the main perceptions of those surveyed about this pillar:

- **Project Manager 3:** "You need to rank and identify the stakeholders for the projects."
- **Project manager 7:** "Consider stakeholders all those who have interest, participation and commitment to project risks, from the coordinator to the project team. Considered that the process is not formalized, although the synergy of the team is compromised so that the movement of all in the project is full and active."

Despite the identified STANDARDIZED level, the company presented being very focused on the development of its year maturity that refers to this item, as seen in the results above.

## CONCLUSION

Based on the application of the proposed model and the results collected, it was observed that the company's current maturity level analyzed in this case study is STANDARDIZED. In some preliminary interviews with some of the project managers, without even being presented to the proposed model, they already pointed the company to an intermediate position in the proposed scale. This can be proven after the application and tabulation of the data. One of the great gains of the application of the model is not only the informative and result of the level of maturity, but also to know its weaknesses, in order to set goals of improvement and training with those involved.

In some companies, after conducting the maturity evaluation, an extensive training program is proposed, in order to better the maturity of the company. We also have to emphasize the great role of human resources within the company, because as we lose mature staff in project management, of course, knowledge is taken away, the famous turn over.

Another important point is, one of the great tools used in project management as a source of elevation of the level of maturity is called training. As project managers and their teams are empowered and updated on existing project management best practices. This allows you to channel knowledge so that it is applied in practice.

As a good practice, always at the end of an evaluation, the negative points are discussed intensively in order to be defined as potential improvement for the organization. Once defined, they are placed in a plan of action that gathers the name of the action, description of what should be done, how long it should be done and who will be responsible. It may be that some actions, after their implantations, require a time of maturation by the company team. This is extremely normal in scenarios where the team with a whole or part of it needs to be empowered in the new knowledge generated through the proposed action. As a product of all this, a small maturity project is created within the organization to guide the company to reach its next level and in how much time. Of course, some actions may have costs or need investments. In these cases, it should be discussed with the top management of the company how they would be handled in case of non-approval or lack of budget for their implementation.

It was observed that the pillar with the lowest maturity evaluation was the Performance and Results pillar whose main objective is to identify tools for monitoring the main results, as well as the progress of the main activities of the project. His maturity level is 2.3. This low level of maturity can be verified through the qualitative data collected through the perceptions of the project managers during the execution of the research. In some cases, some project managers cite that, although the tool is implemented, it is not widely disseminated to the company, which leads some project managers to use it and others do not. It is recommended that the company, when developing or implementing a new project management tool, standardize the application of this tool for all project managers.

As main limiting factors of the research, we observed the following:

- There were few interviewees participating in this research, and for future applications in a company with a project office a greater number of project managers;
- The case study was limited to the application in a mining company, which is not limited to the application of this same questionnaire in companies with similar characteristics.

For a better calibration of the model with the application to the same company, it is suggested, once the chronogram of actions raised after the first evaluation, the re-evaluation of the company again considering the same public previously evaluated. With this, it is expected to compare key breakthroughs, strengths and weaknesses. The process of measuring maturity is a cyclical and dynamic process, that is, one must constantly reevaluate the company, because as mentioned, factors such as the turnover of the teams can influence the maturity of the company.

## REFERENCES

Barreto, M. L. (2001). *Mining and Sustainable Development: challenges for Brazil*. Rio de Janeiro: CETEM/MCT.

BNDES – Banco Nacional do Desenvolvimento. (2010). New steel projects that will increase the steel demand until 2014. *Magazine Vision of Development, 83*. Available: http://www.bndes.gov.br/SiteBNDES/export/sites/default/bndes_pt/Galerias/Arquivos/conhecimento/visao/Visao_83.pdf

Calaes, G. D. (2009). *Comparative Analysis of the Competitiveness of the National Mineral Sector.* MME – Ministério de Minas e Energia. Available: http:// www.mme.gov.br/.../P02_RT06_Anxlise_Comparativa_da_Competitividade_do_Setor_Mineral_Nacional.pdf

DNPM - Departamento Nacional De Produção Mineral. (2006). *Mineral Business: Investor's Guide in Brazil*. Available: http://www.ibram.org.br/150/15001005.asp?ttCD_CHAVE=21653

Ferreira, G. E. (2002). *The Competitiveness of Iron Mining in Brazil*. MCT – Ministério da Ciência e Tecnologia. Available: http:// www.cetem.gov.br/publicacao/series_sed/sed-50.pdf

Germani, D. J. (2002). *The mining in Brazil*. Brasil: Relatório Final.

IBRAM - Instituto Brasileiro De Mineração. (2009). *The industry of mining in Minas Gerais*. Available http://www.ibram.org.br/sites/1300/1382/00000177.pdf

IBRAM - Instituto Brasileiro De Mineração. (2011). *Information and Analysis of the Mineral Economy in Brazil*. Available: http://www.ibram.org.br/

Machado, R. Z., & Fonseca, P. S. M. (2010). *Basic inputs: diagnosis and perspectives*. BNDES – Banco Nacional do Desenvolvimento. Available: http://www.bndes.gov.br/SiteBNDES/export/sites/default/bndes_pt/Galerias/Arquivos/conhecimento/livro_brasil_em_transicao/Brasil_em_transicao_cap20.pdf

Triviños, A. N. S. (1987). *Introduction to research in social sciences: qualitative research in education*. São Paulo: Atlas.

Wagner, A. (2002). *The presidential election and mining*. Gazeta Mercantil.

Yin, R. K. (2001). *Case Study: Planning and methods*. Porto Alegre: Bookman.

# Chapter 11
# Project Maturity Analysis in the Energy Sector


## ABSTRACT

*This chapter will show a macrosystemic view of the energy sector, highlighting its particularities and challenges for the implementation of projects in the sector. The objective of this analysis is to bring enough elements to compose a panorama of the sector in the country, highlighting points in common with organizations related to the theme in the world, allowing establishing justifications for the result found in the maturity research contained in this work.*


## PROBLEMATIZATION

Electricity is considered a basic consumer item, but it carries in its systemic structure a complexity that goes far beyond the availability and consumption of the resource. Companies that project their growth proportionately also need to scale their infrastructure to receive enough energy, otherwise they will face a major constraint in the energy consumption chain that is not trivial to solve.

Countries recognized for being major investors in infrastructure, before dared to project their production at equally bold levels by forcing the division of the globe into groups that also reflect their goals. Groups that were formerly known as first, second, and third world, have seen their classification changed at least four times in the last century.

DOI: 10.4018/978-1-5225-5864-4.ch011

At the date of publication of this book the capitalist vision is the one that dictates the tonic of the relations of consumption in the globe. Launching supply and demand as a mantra for investments and profit margins, for the energy industry is no different.

The bottleneck and free flow vision also applies to the energy market, all sectors are now dependent, from farming to service delivery, considered a growth limiting parameter or one of the agents that most cooperate with increasing efficiency in the production of the market.

What is the level of maturity in the management of projects in an energy company in Minas Gerais? It is based on the central question of this proposed maturity evaluation through which one intends to investigate.

As a secondary question, based on the research carried out in these companies, will be evaluated the impacts created by the maturity obtained in the stakeholders network of the evaluated company.

In the next section, it is reported a profile of the mining sector in the State of Minas Gerais, Brazil.

## THE ENERGY SECTOR

According to CEMIG (2017), one of the main concerns of countries that wish to be recognized for their high capacity is the diversification of energy sources, in particular, the generation of electricity. CEMIG is an important player in Energy Sector in Brazil.

The businesses requirements that are relevant to the industry, can be considered as major responsible for the maturity in the energy industry, are delimited:

- Capability of investment planning based on market growth;
- The allocation of resources for the provision of infrastructure projects;
- The importance of investment in renewable energies;
- Understand the business life cycle considering generation, transmission and distribution;
- The level of compliance with public and market policies.

Besides these important drivers mentioned above, we can highlight the growth in the use of renewable energies. Being regenerative with time, having long cycles of use, paying attention to the maintenance of the means of capture and not to the inexhaustible source of resources:

- Hydro energy
- Solar energy
- Wind Energy
- Biomass
- Solid fuels
- Liquid fuels
- Gaseous fuels
- Geothermal energy
- Tidal Power

In this context, the greater the maturity in the management of structural projects of the sector, the more prepared will be the organization for the use of these energies. The presence of energy sufficiency is not celebrated as a strategic differential for governments or institutions, but its lack turns it into a relentless delimiter for the growth necessary for the market to remain.

## THE APPLICATION AND EVALUATION OF THE PROPOSED CASE

### Data Collection

According to Triviños (1987), one of the main benefits of case study is the deepening of a certain reality. In this case, the results obtained by means of these are valid only for the specific case of the study. According to Yin (2001), the case study brings to the fore the understanding of individual, organizational, social and political aspects.

The evaluation took place through field research, where an energy company was selected, with operations in the Brazilian market. It was decided to evaluate the maturity in project management, following the guidelines of the proposed model. The research work was directed to the responsible project manager, having more than 20 years of experience in the company, was able to respond to the company's project management approach. The field study technique was used, and a semi-structured questionnaire was applied based on the proposed model.

## Selected Case

The selected case is an energy company, being responsible for projects of design of generating units of energy until its conservation. Today the company is a national reference in the subject, placing itself as a technical intermediary between investors and owners, guiding, recommending, investigating and studying the various technical stages of a new enterprise.

## Treatment of Data

The data collection instrument of the proposed model was used, being forwarded in advance to the respondent, and obtained its response collected through interview, as already mentioned above. For this, the personal data of the interviewees as well as the names of the companies in which they execute mining projects were kept.

The data collected by the research were presented in a direct way in this work, and no averages were applied for its presentation. In the end, a score between 0 and 7 was obtained, which represented the individual maturity for each separate pillar, the maturity of the company being the simple arithmetic mean of the nine pillars.

# PRESENTATION OF RESEARCH RESULTS

## Profile of Respondents

### Analysis of Quantitative Research Data

The research was sent to the project managers by the company's e-mail with a deadline for confirmation of the data through the interview with the indicated professional.

For the case study, the final maturity level was 3.35, which, observing the scale presented in the table below, shows that the company presents itself at a STANDARDIZED level of maturity.

#### *Final Score by Maturity Level*

For the STANDARDIZED maturity level, this level represents the state enterprise as:

*Table 1. Pillar maturity evaluation*

| Item | Pillar | Total |
|---|---|---|
| 1 | Project Management Office | 5. 50 |
| 2 | Project Management | 4.12 |
| 3 | People Management | 3.89 |
| 4 | Methods and Practices | 3.50 |
| 5 | Organizational Structure | 3.50 |
| 6 | Performance and Results | 2.33 |
| 7 | Engineering and Geology | 4.20 |
| 8 | Environment Management | 0 |
| 9 | Stakeholder Management | 3.11 |
| | **MATURITY** | **3.35** |

Source: Author's proposal

*The company is already promoting standardization of a methodology in the management of mining projects, including the process of training hierarchical levels. At this level the company begins the process of restructuring and adaptation of the structure to attend the management of the projects, we start thinking about the implementation of a PMO - Office of Projects as well as management software. (Proposal of the author)*

In the chart below we can observe the maturity of each pillar, as well as the final maturity for the case study studied and its respective placement within the maturity scale.

## Qualitative Research Data Analysis

In addition to the quantitative data presented in the previous section, the research gathered qualitative data by collecting the project manager's impressions on each observed pillar. The data will be presented for each of the nine pillars observed.

The result found is a cross-sectional photograph of the mining industry for evaluation. By evaluating the results obtained in each pillar, it is possible to establish investment strategies such as deployments, acquisitions or even mergers.

In the following sections, we will do an analysis of each pillar of the research. It brings, in addition to the statistical result, a context view analyzed

*Figure 1. Overall maturity rating per pillar*
*Source: Research data*

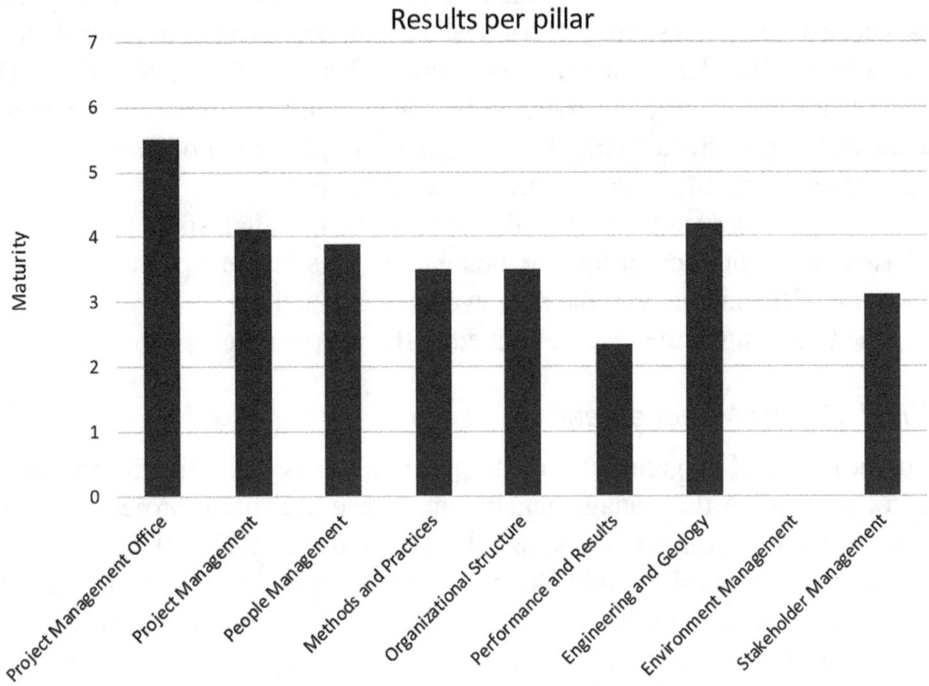


longitudinally. We seek to find cause and effect relationships, groupings and risks for investment in the energy sector.

## *Pillar: PMO - Project Management Office*

The PMO pillar - Project Management Office – has had as its main objective to collect information on the existence or presence of a project office within the research company. Regarding the level of maturity, this pillar received the grade 5.5 according to research, that is, it is currently at the IMPLEMENTED level.

The main justification presented for this level of maturity was the moment in which the company is, with respect to the implementation of procedures and its adequacy to the practices of the company and of the market that it serves.

In this way, it is observed, through the data collected above, that the project office already has tools for the management of projects in use, according to the definition of maturity level IMPLEMENTED, characterized for this item.

The following is the result obtained in the project management pillar.

### Pillar: Project Management

With respect to the Project Management pillar, its objective is to identify the presence of tools that support and assist in the management of projects. In the case of this pillar, the maturity assessed was 4.12, also on the CONTROLLED scale. One of the main considerations of this level of maturity is the commitment with the standardized methodology, having implanted processes such as change management, lessons learned and indicators.

It can be observed from the collected information that, some tools are in phase of consolidation of the use, being one of the main responsible for the increase of the maturity in the next cycles of evaluation.

The following is the result obtained in the people management pillar.

### Pillar: People Management

For the People Management pillar, its main function is to identify the presence of tools that allow the management of human resources of the projects. In this case, its maturity evaluation was 3.89, standing on the STANDARDIZED scale.

In this case, we highlight the existence of an active human resources sector, guiding each professional about their job functions and instructions. As points of improvement, the lack of use of a formal project manager career development tool and a performance management tool for the team. The follow-up meetings are usually held with a fixed schedule, allowing the team to prepare the necessary information for decision making.

The presence of an organization chart in the company transmits visibility to the employees, with respect to the necessary report and the escalation of decisions in the company.

The focus of actions of people management should be in generating an environment of trust in the organization. Where improvement points are indicated, they must generate a sufficient number of actions and stakeholders to be successful. Inevitably, people should be included in the center of this context, with sufficient autonomy and contextualization to carry out their work.

Without the patronage of the people, impossible to carry forward actions of any other pillar.

The following is the result obtained in the methods and practices pillar.

### Pillar: Methods and Practices

The main objective of the Methods and Practices pillar is to identify the existence of a well-defined flow of processes to assist project managers in the execution and conduct of their projects. The maturity level evaluated for this pillar was 3.50, also at the STANDARDIZED level.

As a characteristic of the company with respect to the pillar collected, we can highlight the feasibility study that is carried out before the beginning of the project, benchmarking practices being carried out in similar companies, the practice of using cost control through estimation of added value and a formal process of closing contracts open during the project.

In this way, it is observed that the tools being developed by the project office and applied by project managers are still in the process of being consolidated to become part of the company's project management process.

The following is the result obtained in the structure pillar.

### Pillar: Structure

For the Structure pillar, we seek to identify the presence of an adequate organizational structure for the management of the company's projects. In this case, the maturity level evaluated was 3.50, which leads us to understand that the company is in the structuring phase to suit its structure format.

The following characteristics can be considered for the identified result: The designation of a project manager by the organization, the role of Project Manager established by the company and the wide communication in the organization of the beginning of a new project, where the designation of professionals will take place which will be allocated on demand.

The following is the result obtained in the performance and results pillar.

### Pillar: Performance and Results

In the case of the Performance and Results pillar, its main objective is to identify tools for monitoring the main results, as well as the progress of the main project activities. Its level of maturity is in 2.33, that is, in a SCHEDULED level. Highlighting the fulfillment of the requirements established by the customer and the customer satisfaction of the project with respect to the developed product, it presents a counterpoint with respect to the other objectives such as the fulfillment of deadlines, costs and evaluation of the project team, which did not obtain a satisfactory level in the evaluation performed.

This was the pillar with the second lowest maturity assessment. It demonstrates among other characteristics the need to use the tools of management control or the implementation of systems that can guarantee the management of performance and results. It is one of the main drivers of efficiency for the governing body of the organization and the indicator of commitment of the project stakeholders with the goals established by the leaders.

The following is the result obtained in the engineering and geology pillar.

### Pillar: Engineering and Geology

The Engineering and Geology pillar aims to identify if the organization is following the proposed process of engineering phases aligned to the life cycle for the geological investigations. Its maturity rating stood at 4.2, level CONTROLLED. The collected data present an established process, where there is indication that there are controls in the projects and the establishment of a master plan that aims to guide the bases for the execution of the indicated demands.

The projects being run by the organization respect the flow of composition, following the steps of conceptual engineering, basic and detailed. In addition to the exploratory drilling being an input for conceptual engineering, the research demonstrates the existence of important milestones for the performance of the abutment pillar.

As points of improvement, 3D engineering is still not applied as a technology that makes up the technical solution, besides the non-use of VIP (value improving practices) to increase the performance of the project.

The following is the result obtained in the environmental management pillar.

### Pillar: Environmental Management

Not punctuated, since it does not apply to the market that company operates.

### Pillar: Stakeholder Management

The main objective of the Stakeholder Management pillar is to identify the presence of tools and processes that assist the company in the identification and knowledge of all stakeholders that are affected positively or negatively by the project. For this case, the maturity stood at 3.11, its level being identified as STANDARDIZED.

Despite the level identified as STANDARDIZED, there are important points of improvement that point to the correct management of stakeholders and the achievement of expectations as one of the most important strategies of project management in the organization.

The author's conclusion regarding the research carried out is presented below.

## CONCLUSION

According to the proposed model and the results achieved in the maturity research, it can be observed that the company has a STANDARDIZED maturity level: highlighting the application and use of good practices that add value to the project management and engineering process; bringing more security and fewer project changes throughout the life cycle.

Among the proposed improvements, we can highlight:

- The application of recurrent maturity surveys;
- The realization of an improvement plan considering the 9 pillars in the research;
- Monitoring compliance with the improvement plan that may contain training, guidance and consulting;
- The recognition of professionals who cooperate with the increase of the maturity of the company with the fulfillment of the established goals;
- A gradual approach to project management at the company's C-level, as positive results are achieved;
- Value added to the product and project provided by the company due to the increase in excellence in its projects.

The maintenance of a monitoring and control agenda as mentioned previously, it is more natural to direct investments in actions that aim to increase efficiency in the realization of the company's projects.

Much of what is expected with respect to the security of investments in this sector can be obtained through maturity research like this one, allied to action plans / continuous improvement projects. Through continuous cycles of evaluation - action, it is possible to perceive the increase of the maturity of companies in this sector, taking the maturity indices to levels always higher than the previous research.

The STANDARDIZED result obtained presents a major challenge with respect to investment safety, but also offers a broad horizon of evolutionary strategies. The flexibility in the definition of actions tends to decrease as the company becomes mature.

One of the great challenges for the company studied is the maintenance of its plan of action and projects assumed after the maturity research. Efficient initiatives, at the same time not so popular can be assumed by the organization's executives. When deployed in a recurring manner, it generates a cycle of improvement and recurrent evaluation in which a better result in relation to the previous research, motivates its leaderships and brings back the perception that they are on the right track.

On the other hand, actions that have been compromised and not executed can generate a setback that tended to undermine the company environment. Creating waves of discredit on the part of its collaborators and also in the investors. Actions such as reviewing improvement strategies should be applied repeatedly in these companies to regain confidence and generate new opportunities.

## REFERENCES

CEMIG – Companhia Energética De Minas Gerais. (2017). Available: http://www.cemig.com.br/pt-br/a_cemig/nossos_negocios/Energia/Paginas/default.aspx

Triviños, A. N. S. (1987). *Introduction to research in social sciences: qualitative research in education.* São Paulo: Atlas.

Yin, R. K. (2001). *Case Study: Planning and methods.* Porto Alegre: Bookman.

# Chapter 12
# Project Maturity Analysis in Civil Construction


## ABSTRACT

*This chapter applies a macroview of civil construction, punctuating its particularities and challenges for the implementation of projects in the country. The objective of this analysis is to provide sufficient elements to compose a panorama of the civil construction sector in the country, highlighting points in common with organizations related to the theme in the world, allowing establishing justifications for the result found in the maturity research contained in this work.*


## PROBLEMATIZATION

Currently, construction is responsible for a significant portion of the country's gross domestic product (GDO). In addition to being an important indicator of a country's development, civil construction today is impacted by political crises and the downturn in the economy, caused by any of the factors of demotivation in investment.

In addition to the challenges caused by the slowdown in investments in structuring projects, the sector faces challenges such as the productivity of its processes, the rational use of its resources and the adaptation to environmental management policies, increasingly recognized for their rigor and strong engagement of society.

DOI: 10.4018/978-1-5225-5864-4.ch012

The waste minimization, oil crises, and energy efficiency are issues that directly impact the way projects are managed in the industry. Requiring from the organizations involved a knowledge that, historically, was only found in the sectors responsible for each of these issues, it now requires a holistic view of the enterprise, permeating the technical issues, important for the good progress of the project, but also areas such as sustainability and social responsibility.

In this context, then, it is based the central question of this proposed maturity evaluation through which one intends to investigate: **What is the level of maturity in project management in a construction company in Minas Gerais?**

As a complementary question, based on the research carried out in this company, will be evaluated the impacts generated by the maturity obtained in the network of stakeholders of the evaluated company. How does this interfere in the chain of sponsorship and trust of employees and investors?

In the next section, it is reported a profile of the civil construction sector in the State of Minas Gerais, Brazil.

## THE CIVIL CONSTRUCTION SECTOR

According to CBIC (2017), Brazil experienced a lower investment for construction in the current period, with less construction in progress due to the decrease of GDP.

For the organizations and nations that want to be recognized for the legacy of infrastructure left for future generations, it is imperative to consider issues that go beyond constructability, pointing to concerns expressed in the VIP (value improvement practices) that address a better use of natural resources, but also the increase excellence in the execution of the sector's.

The way of delimiting the business requirements that are relevant to the sector, it can be considered a major responsible for the maturity in the construction industry:

- Capability of investment planning based on market growth;
- The allocation of resources for the provision of infrastructure projects;
- The importance of investment in renewable raw materials;
- The level of compliance with public and market policies;

In addition to these important drivers, we can highlight the growth in the use of renewable resources.

In this context, the greater the maturity of project management in the sector, the more prepared the organization will be to face the challenges of sustainability, competition and financial results.

## THE APPLICATION AND EVALUATION OF THE PROPOSED CASE

### Data Collect

According to Triviños (1987), one of the main benefits of case study is the deepening of a certain reality. In this case, the results obtained by means of these are valid only for the specific case of the study. According to Yin (2001), the case study brings to the fore the understanding of individual, organizational, social and political aspects.

The evaluation was carried out through field research, where a commercial infrastructure construction company was selected. It was decided to evaluate the maturity in project management, following the guidelines of the proposed model. The research work was directed to the responsible project manager, having wide experience in project management in the sector, was able to respond to the enterprise management approach of the company. The field study technique was used, and a semi-structured questionnaire was applied based on the proposed model.

### Selected Case

The selected case is a construction company from Minas Gerais, founded in 1995, responsible for projects with photo in incorporation and construction, also acting in the real estate, financial and hotel management. The company has in its portfolio projects executed in Minas Gerais and Rio de Janeiro, possessing as a premium awards received since its inauguration.

### Treatment of Data

It was used the data collection instrument of the proposed model, being forwarded in advance to the respondent, obtained his answer collected through interview. For this, the personal data of the interviewees as well as the names of the companies to which the model was applied were stored.

The data of the research were collected and presented in a direct way in this work, not applying averages for its presentation. In the end, a score between 0 and 7 was obtained, which represented the individual maturity for each separate pillar, the maturity of the company being the simple arithmetic mean of the nine pillars.

## PRESENTATION OF RESEARCH RESULTS

### Profile of Respondents

#### Analysis of Quantitative Research Data

The research was sent to the project manager by e-mail, with a deadline for confirming the data through the interview with the indicated professional.

For the case study, the final maturity level was 4.33, which, observing the scale presented in the table above, shows that the company presents a maturity CONTROLLED level.

For the CONTROLLED maturity level, this level represents the state enterprise as:

*Table 1. Pillar maturity evaluation*

| Item | Pillar | Total |
|---|---|---|
| 1 | Project Management Office | 6.00 |
| 2 | Project Management | 3.71 |
| 3 | People Management | 2.33 |
| 4 | Methods and Practices | 3.50 |
| 5 | Organizational Structure | 3.50 |
| 6 | Performance and Results | 5.83 |
| 7 | Engineering and Geology | 4.76 |
| 8 | Environment Management | 7.00 |
| 9 | Stakeholder Management | 2.33 |
| | **MATURITY** | **4.33** |

Source: Author's proposal.

*The company's senior management is already committed to the need for a standardized project management methodology; all processes are implemented such as: change management, lessons learned, evaluation and management by indicators, trained managers ahead of mining projects. In short, there is already a widespread culture. (Proposal of the author)*

In the chart below we can observe the maturity of each pillar, as well as the final maturity for the case study studied and its respective placement within the maturity scale.

## Data Analysis: Qualitative Research

In addition to the quantitative data presented in the previous section, the research gathered qualitative data by collecting the project manager's impressions on each observed pillar. The data will be presented for each of the nine pillars observed.

*Figure 1. Overall maturity rating per pillar*
*Source: Research data.*

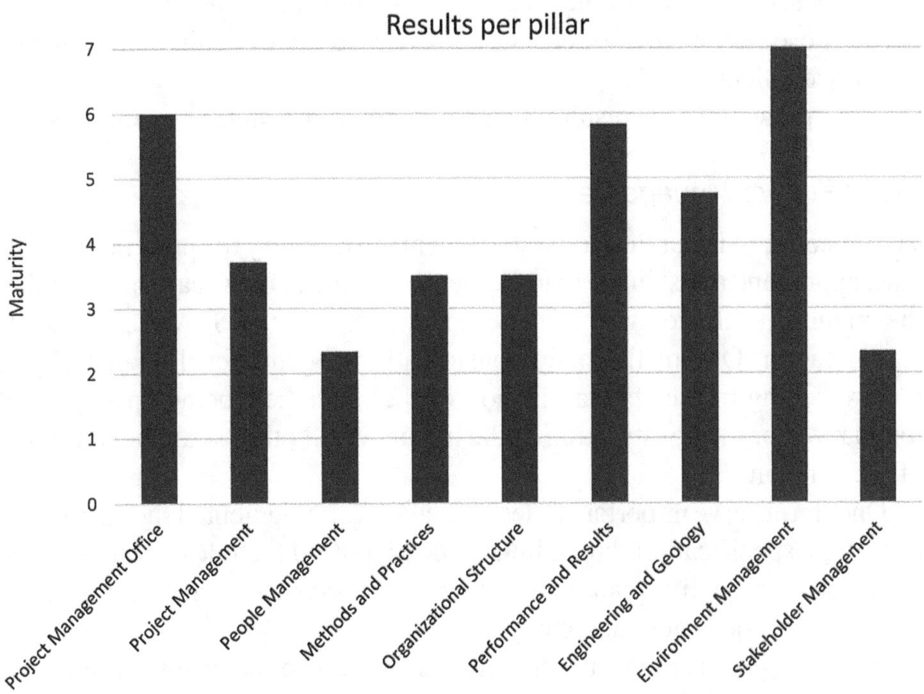

The result found is a cross-sectional photograph of the mining industry for evaluation. Through the evaluation of the results obtained in each pillar, it is possible to establish investment strategies such as deployments, acquisitions or even mergers with companies of the same sector.

In the following sections, we will do an analysis of each pillar of the research. It brings, in addition to the statistical result, a context view analyzed longitudinally. It seeks to find, cause and effect relationships, groupings and risks for investment in the construction sector.

## Pillar: PMO - Project Management Office

The PMO pillar - Project Management Office - had as main objective, to collect information about the existence and performance of a project office within the company surveyed, regarding the level of maturity. This pillar received the grade of 6.0 according to research, that is, it is currently at the STRATEGIC level.

The main justification presented for this level of maturity was the positioning of the business of the company, being mandatory to the PMO the link of the projects to the strategic planning of the company. This development reaches the indicators of projects, being an approach that increases the perception of value with the client of the enterprise.

It is observed, through the data collected above, that the project office is one of those responsible for ensuring the perception of quality and value by the project client.

The following is the result obtained in the project management pillar.

## Pillar: Project Management

The objective of pillar Project Management is to identify the presence of tools that support and assist in the management of projects. In the case of this pillar, the maturity evaluated was 3.71, being classified in the STANDARDIZED classification. One of the main considerations of this level of maturity is the awakening to the standardization of the processes, being an important foundation for the construction of the maturity and other strategic unfolding of the company.

One can observe important milestones already implemented such as: the use of a scope specification with a link in the identified technical requirements, existence of a quality plan and standard processes for contracting such as engineering, equipment and etc.

The following is the result obtained in the people management pillar.

## Pillar: People Management

For the People Management pillar, its main function is to identify the presence of tools that allow the management of human resources of the projects. In this case, its maturity assessment was 2.33, standing on the SCHEDULE scale.

For this case, it highlights basic initiatives, but important for increasing the maturity of the company in the aspect of people management. It is possible to highlight the existence and use of a plan of team allocation and delegation of authority, making the team feel participant in decisions, but at the same time endowed with limited autonomy, being enough for the decision-making process expected by the company.

Another important aspect is the formal request for resources to be submitted to the HR before the start of the project, allowing the participation of each party at the right time. It is also valid for the demobilization of resources, without prejudice to the stakeholders involved.

The focus of actions of people management should be in generating an environment of trust in the organization. Where improvement points are indicated, they must generate a sufficient number of actions and stakeholders to be successful. Inevitably, people should be included at the center of this context, with sufficient autonomy and contextualization to carry out their work.

Without the patronage of the people, is impossible to carry forward actions of any other pillar.

The following is the result obtained in the methods and practices pillar.

## Pilar: Methods and Practices

The main objective of the Methods and Practices pillar is to identify the existence of a well-defined flow of processes to assist project managers in the execution and conduct of their projects. The maturity level evaluated for this pillar was 3.50, also at the STANDARDIZED level.

The characteristics of the company with respect to the pillar collected, we can highlight the feasibility study that is carried out before the beginning of the project, benchmarking practices being carried out in similar companies, the practice of using cost control through estimation of value added and a formal process of closing contracts open during the project.

It is observed that the tools being developed by the project office and applied by project managers are still in the process of being consolidated to become part of the company's project management process.

The following is the result obtained in the structure pillar.

### Pillar: Structure

For the structure pillar, it is sought to identify the presence of an adequate organizational structure for the management of the company's projects. In this case, the maturity level evaluated was 3.50, which leads us to understand that the company is in the structuring phase to suit its structure format.

The following characteristics can be considered for the identified result: 1. The designation of a project manager by the organization, having the function compatible with similar positions as manager of human resources, manager of operations, etc. 2. The formal communication about the beginning of the project activities, being formally authorized to the project sponsor and executives co-responsible for its feasibility.

The following is the result obtained in the performance and results pillar.

### Pillar: Performance and Results

In the case of the Performance and Results pillar, its main objective is to identify tools for monitoring the main results, as well as the progress of the main project activities. Its level of maturity is in 5.83, that is, at an IMPLEMENTED level: highlighing the implementation of a unique methodology of projects and use adherence throughout the executing organization; existence of a PMO that supports the corporate management of projects. People are trained and know the methodology and process of project management for projects.

The pillar that presented a high maturity in the research, demonstrating an advance regarding the management of project deadlines, the customer satisfaction with the product delivered to the end of the project, the assertiveness of the project budgeting, besides the effective monitoring of the satisfaction of the project, the lifecycle of the project. It is also worth noting the use of the evaluation process of the project manager during the project.

The following is the result obtained in the engineering and geology pillar.

### Pillar: Engineering and Geology

The Engineering and Geology pillar aims to identify if the organization is following the proposed process of engineering phases aligned to the life cycle for the geological investigations. Its maturity rating stood at 4.76, level CONTROLLED. The collected data present an established process, where there is indication that controls in the projects and the establishment of a master plan that aims to guide the bases for the execution of the indicated demands.

The projects being run by the organization respect the flow of composition, following the steps of conceptual engineering, basic and detailed. In addition to the exploratory drilling being an input for conceptual engineering, the research demonstrates the existence of important milestones for the performance of the abutment pillar.

The 3D engineering is not yet applied as a technology that makes up the technical solution, besides the non-use of VIP (value improving practices) to increase the performance of the project.

The following is the result obtained in the environmental management pillar.

### Pillar: Environmental Management

In the Environmental Management pillar, the company showed its highest maturity, registering an index of 7.0. In line with the most advanced in environmental management, the company places itself as a reference in the consideration, application and monitoring of parameters that indicate a high maturity in this requirement.

It is worth mentioning the formalization of environmental studies at the beginning of the project, considering the mapping of the biotypes involved (areas of environmental impact), also guaranteeing the process of surface mapping impacted by the project during environmental licensing and conceptual engineering. The alignment between engineering development and environmental activities is also a relevant factor in maturity research.

As a differential of the area, there is a formal process to monitor the pre-deployment release and operating licenses with the government agencies involved in the venture.

The following is the result obtained in the stakeholder management pillar.

### Pillar: Stakeholder Management

The main objective of the Stakeholder Management pillar is to identify the presence of tools and processes that assist the company in the identification and knowledge of all stakeholders that are affected positively or negatively by the project. For this case, the maturity stood at 2.33, its level being identified as SCHEDULED.

At this level, it is important to highlight the classification of groups of stakeholders, facilitating the creation of strategies of approach and differentiated involvement for each public of the project. Centralized stakeholder contact data also helps with this index, thus ensuring easy mobilization as unexpected situations occur in the project. In addition to identification, the content of the well-defined communication is highlighted, relying on the periodicity in which these should occur, which communication medium is used and which group of stakeholders these should interact with.

The author's conclusion regarding the research carried out is presented below.

## CONCLUSION

According to the proposed model and the results achieved in the maturity survey, it can be observed that the company has a maturity level CONTROLLED. Emphasizing the control of good practices already applied to the company's projects, allowing for greater performance and fewer project changes during the implementation life cycle.

Among the proposed improvements, we can highlight:

- The application of recurrent maturity surveys;
- Creation of a unified project and change management methodology, in order to guarantee the adhesion of the organization's stakeholders;
- Creation of indicators to ensure the monitoring of processes that have already been implemented;
- The recycling of knowledge through training, guidance and consulting;
- The recognition of professionals who cooperate with the increase of the maturity of the company with the fulfillment of the established goals.

With the maintenance of a monitoring and control agenda as mentioned previously, it is more natural to direct investments in actions that aim to increase efficiency in the company's projects.

Much of what is expected with respect to the security of investments in this sector can be obtained through maturity research like this one, allied to action plans / continuous improvement projects. Through continuous cycles of evaluation - action, it is possible to perceive the increase of the maturity of companies in this sector, taking the maturity indices to levels always higher than the previous research.

The IMPLEMENTED result obtained presents a major challenge with respect to growth stabilization. The flexibility in the definition of actions tends to decrease as the company becomes mature.

One of the great challenges for the company studied is the consolidation of its action plan and projects assumed after the maturity research. These cycles of research and improvements when deployed in a recurring manner, generate recurring improvements in results compared to the previous research and motivates their leaders and brings the perception that they are on the right track.

On the other hand, actions that have been compromised and not executed can generate a setback that tends to undermine the company environment, creating waves of discredit on the part of its collaborators and also in the investors. Actions such as reviewing improvement strategies should be applied repeatedly in these companies to regain confidence and generate new opportunities.

## REFERENCES

CBIC – Câmara Brasileira da Indústria da Construção. (2017). Available: http://www.cbicdados.com.br/menu/pib-e-investimento/pib-brasil-e-construcao-civil

Triviños, A. N. S. (1987). *Introduction to research in social sciences: qualitative research in education*. São Paulo: Atlas.

Yin, R. K. (2001). *Case Study: Planning and methods*. Porto Alegre: Bookman.

# Appendix

## RESUME OF THE MATURITY ASSESSMENT IN MINING, ENERGY, AND CIVIL CONSTRUCTION SECTORS

### Mining Sector

The research was based on the Project Management Maturity study of a large iron ore mining company, concentrated in the state of Minas Gerais, and which has one of the largest investment budgets for expansion projects for the next five years. As informed before, the project manager has been named by number in order to respect each one of them. Find in the sequence the assessment of the 09 projects managers interviewed during the maturity evaluation.

#### Project Manager 1

See Tables 1-9.

#### Project Manager 2

See Tables 10-18.

#### Project Manager 3

See Tables 19-27.

## Appendix

*Table 1. Project management*

| Id. | Question | Score |
|---|---|---|
| **01** | Is there any document to formalize the existence of the project in your area? | 7 |
| **02** | Does the project existence document adequately describe the project output? | 7 |
| **03** | Is there a document that can document the assumptions (prerequisites necessary for you to complete the work) and constraints (dates imposed by the client, limitation of funding, limitation of resources) of the project? | 7 |
| **04** | Is there a Project Analytical Framework (EAP) or WBS? | 7 |
| **05** | Is there any tool to formalize the client's acceptance? | 0 |
| **06** | Is there a baseline against which project progress, costs and deadlines are measured? | 0 |
| **07** | Is there an established process for project change control? | 0 |
| **08** | Is there a process of identifying predecessors and successors of project activities? | 0 |
| **09** | Is there a separation of project activities according to each schedule of the resource to be used (business days, holidays, etc.)? | 0 |
| **10** | Is there a process for identifying and prioritizing the critical path of the project? | 0 |
| **11** | Is there a baseline recording of the project at the end of the schedule development? | 0 |
| **12** | In developing the schedule, do all activities have resources? | 0 |
| **13** | Is there a formalization process of generating provisions (contingencies) for project costs? | 0 |
| **14** | Is the process of formalizing a matrix of responsibilities for the project? | 0 |
| **15** | Is there a formalized process of generating and transmitting project information during its execution (Communications Plan)? | 7 |
| **16** | Is there a mapping that details who needs the information, when it will be needed, how it should be defined, and by whom? | 0 |
| **17** | Is there a place with phones, addresses, e-mails and contacts from the entire team and involved? | 7 |
| **18** | Is there a formalized process for defining the schedule of project meetings? | 7 |
| **20** | Are the meetings pre-scheduled from the start and on the agenda of the entire team? | 7 |
| **21** | Does each team member know what information to bring to the follow-up meeting? | 7 |
| **22** | Is there a formal project budgeting process by grouping estimates of activities, resources, hiring, contingencies, etc.? | 0 |
| **23** | Is there a formalized risk identification process for the project? | 0 |
| **24** | Is there a qualitative and quantitative project risk assessment process? | 0 |
| **25** | Is there a project risk prioritization process? | 0 |
| **26** | Is there a process for planning the project's risk responses? | 0 |
| **27** | Is there a process to identify who is responsible for the project risk? | 0 |
| **28** | Is there a risk management and monitoring tool for the project? | 0 |
| **29** | Is there a process for identifying project product requirements? | 0 |
| **30** | Are the project product requirements translated into the work required to develop this product? | 0 |
| **31** | Is there a document that can formalize the work of the project (Scope Statement)? | 7 |
| **32** | Is there a formalized audit plan for quality requirements for the project (Quality Plan)? | 0 |
| **33** | Is there formalized planning for the provision and contracting of project resources (engineering, equipment, etc.)? | 0 |
| **34** | Is there a formalized contracting cycle for large project deliverables (Bid Cycle or Hiring Cycle)? | 0 |

*Table 2. Performance and results*

| Id. | Quesion | Score |
|---|---|---|
| 01 | Are deadlines met as planned? | 0 |
| 02 | Are the quality of the delivered projects according to the customer's requirements? | 7 |
| 03 | Are project costs met as planned? | 0 |
| 04 | Is there a monitoring of customer satisfaction with the project product? | 0 |
| 05 | Is there a process for evaluating the performance of the project manager in the conduct of the project? | 0 |
| 06 | Is there a project team evaluation process? | 7 |

*Table 3. Methods and practices*

| Id. | Question | Score |
|---|---|---|
| 01 | Before the start of the project is a feasibility study for the investment carried out? | 7 |
| 02 | Is there a practice of benchmarking projects in relation to projects developed by similar companies? | 0 |
| 03 | Is there practice of contracting consultancies to identify and evaluate project maturity for successful projects (IPA - Independent Project Analysis, etc.)? | 0 |
| 04 | Is there practice of hiring consultancies in the evaluation and optimization of budget? | 7 |
| 05 | Is there a formal process of closing contracts open during the project? | 0 |
| 06 | Is there a formal process of collecting the lessons learned during and at the end of the project? | 0 |
| 07 | Is there a formal process of cost control through tools like EVA - Value Added Estimation? | 7 |
| 08 | Is there a formal operational risk analysis and survey process (HAZOP - Hazard and Operability Studies)? | 7 |

## Project Manager 4

See Tables 28-36.

## Project Manager 5

See Tables 37-45.

## Project Manager 6

See Tables 46-54.

***Appendix***

## *Table 4. Engineering and geology*

| Id. | Question | Score |
|---|---|---|
| 01 | Is there a formalized process that every project developed in the organization should go through the phases: conceptual, basic and detailed? | 7 |
| 02 | Is every project being run by the organization going through or going through the conceptual, basic, and detailed engineering phases? | 0 |
| 03 | Is there a formalized process for the drilling and auditing of mineral resources for engineering? | 0 |
| 04 | Is there a formal process that all conceptual engineering should begin with the first results of exploratory surveys? | 7 |
| 05 | Is there a formalization and submission of conceptual engineering information to the Environmental area for environmental licensing? | 7 |
| 06 | Is all Conceptual Engineering developed based on a mapping of biotopes (areas impacted by the environment) and surface? | 7 |
| 07 | Are all process tests (Technological Development) formally sent to Engineering to start Basic Engineering? | 0 |
| 08 | Is there a formalized process for reviewing engineering quality (Peer Review) before moving from one phase to another (eg: Conceptual Engineering for Basic Engineering)? | 0 |
| 09 | Is there a formalized engineering development culture in 3D technology? | 0 |
| 10 | Is there a process of evaluation, prior to the beginning of the Detailed Engineering, on the detailing of the project through the modularization of the construction? (Construction in Modules) | 7 |
| 11 | Is there the process of generating a Master Plan for the project after Conceptual Engineering? | 7 |
| 12 | If yes for answer 11, is there a formalized process of freezing this master plan and communicating it to all project stakeholders? | 0 |
| 13 | Is there a formalized process of generation of Descriptive Memorials, Spreadsheets, Data Sheet and Technical Specifications for the Supplies area? | 7 |
| 14 | Is there a formalized process of participation of the Supplies area during the evolution of the Basic Engineering for Detailed Engineering to determine the type of contracting of the Supply, Assembly and Implementation of the project? | 0 |
| 15 | Is there a formal process of realizing and studying tradeoffs for the lowest cost for utilities (water, gas, electricity, logistics, etc.)? | 7 |
| 16 | Is there a formal process of realization and study of opportunities (tradeoffs) regarding the location of the project (surface workers, communities, logistics, etc.)? | 7 |
| 17 | Is there a formal process for evaluating and validating each engineering phase (Conceptual, Basic, and Detailed) before moving from one to the other (Front End Loading Portions of Independent Project Analysis)? | 0 |
| 18 | At the end of the Project's Conceptual Engineering, is there a formal process to generate a Prefeasibility Study? | 0 |
| 19 | At the end of the Basic Engineering of the Project, is there a formal process to generate a Feasibility Study of the project? | 0 |
| 20 | During the execution of the Conceptual Engineering, is there a formal process of identification and first contacts with impacted landowners as well as local communities? | 7 |
| 21 | During the execution of Conceptual Engineering, is there a formal topographic survey of the region as well as the generation of orthophotos for the project's master plan? | 7 |
| 22 | After Conceptual Engineering, is there a formal process of generating investment estimates at a precision level of -15% to + 25%? | 7 |
| 23 | After the Basic Engineering, is there a formal process of generating the investment estimates at a precision level of -10% to + 10%? | 7 |
| 24 | Is there a formal closed-end planning process for the project after its useful life? | 0 |
| 25 | Is there a formal process for identifying VIPs - Value Improving Practices? | 7 |

*Table 5. Organizational structure*

| Id. | Question | Score |
|---|---|---|
| 01 | When a project starts, is there a project manager assigned by the organization to lead the project? | 7 |
| 02 | Is there an area focused on project planning? | 7 |
| 03 | Is there a formalized position of project manager within the organization? | 7 |
| 04 | Is there formalization of the Project Manager's power to similar functional positions (HR Manager, Operations Manager, etc.)? | 0 |
| 05 | Is there formalization of areas within Project manangement such as: Planning, PMO, Cost Control, etc.? | 7 |
| 06 | Are all internal areas of the company formally communicated from the beginning of the project and are their representatives formally assigned to the project? | 7 |

*Table 6. People management*

| Id. | Question | Score |
|---|---|---|
| 01 | Is there a formal process of developing the project manager's career and team? | 0 |
| 02 | Do follow-up meetings have a schedule and a fixed format? | 7 |
| 03 | Does each project member receive their work with instructions and in advance? | 0 |
| 04 | Is there a formal process of variable compensation for project team performance? | 7 |
| 05 | Is there a formal process for evaluating project team performance? | 7 |
| 06 | Is there a formal organization chart of the project team, delegations, etc.? | 0 |
| 07 | Are all the human resources required for the project formally sent to the HR area to arrange recruitment? | 0 |
| 08 | Are all interpersonal skills of the project team managed during the project? | 0 |
| 09 | Are all conflicts between the project team formally addressed? | 7 |

## Project Manager 7

See Tables 55-63.

## Project Manager 8

See Tables 64-72.

## Project Manager 9

See Tables 73-81.

***Appendix***

*Table7. Project manangement office*

| Id. | Question | Score |
|---|---|---|
| 01 | Is there a flow for the arrival in your area of the information about the existence of a project or a contract? | 7 |
| 02 | Is there a dedication of the professionals to the management of the projects? | 7 |
| 03 | Is there any tool for estimating effort (time) and project costs? | 0 |
| 04 | Is there any tool for estimating project resources? | 7 |
| 05 | Is there any tool to formalize the client's acceptance? | 0 |
| 06 | Is there an established process for formal acceptance of each deliverable of the project? | 7 |
| 07 | For the development of project documentation, is there a repository of past project templates for reuse in new projects? | 0 |
| 08 | Is there an area focused on the physical and financial control of the project? | 7 |
| 09 | Is there a visibility room, a mural or even an intranet site where everyone can access projects in a simple way? | 7 |
| 10 | In the case of geographically dispersed teams, are the procedures / technologies used to integrate them adequate? | 0 |
| 11 | Are there managerial reports with a previously defined format agreed upon with the client? | 7 |
| 12 | Is there a formalized process for the closing of a project / contract? | 0 |
| 13 | Is there a formal process for the provision of information in the closing phase (finalization of the work)? | 0 |
| 14 | There is an area focused on the control and support of the project (PMO or Project Office) | 7 |

*Table 8. Stakeholder management*

| Id. | Question | Score |
|---|---|---|
| 01 | Is there a formalized process of mapping and managing the company's internal stakeholders? | 0 |
| 02 | Is there a formalized process for managing stakeholder expectations of the project during its execution? | 7 |
| 03 | Are formal interviews conducted to identify project stakeholders? | 0 |
| 04 | Are all potential impacts caused by each stakeholder formally assessed and managed during and after the project? | 7 |
| 05 | Is there a formal process of ranking and prioritizing the degree of influence of each stakeholder on the project? | 0 |
| 06 | Is there a formalized process of classification of stakeholder groups? | 7 |
| 07 | Is there a formal process of simulating the reactions of each stakeholder to the various situations that may occur during and after the project? | 0 |
| 08 | Is there a formalized process of gathering and filing the information of each stakeholder such as: name, address, telephone, etc.? | 7 |
| 09 | Is there a formalized Communication Plan capable of identifying what kind of information, when, how and for which stakeholder to send? | 0 |

*Table 9. Environment management*

| Id. | Question | Score |
|---|---|---|
| 01 | Is there a formalized process of beginning environmental studies with the beginning of Conceptual Engineering? | 7 |
| 02 | Is there a formalized process for mapping biotopes (areas of environmental impact) to aid in the development of Conceptual Engineering? | 7 |
| 03 | Is there a formalization process of mapping the surface properties impacted by the project during the Environmental Licensing and Conceptual Engineering? | 7 |
| 04 | Is there a formal process of generating documentation for Environmental Impact Studies and Environmental Impact Report by the Engineering area? | 7 |
| 05 | Is there a formal process that synchronizes the development of Engineering and Environment activities? (Generation of Documentation for Environmental Licensing)? | 7 |
| 06 | Is there a formalized area for institutional relations with government agencies? | 7 |
| 07 | Is there a formalized area for relationship with the communities impacted by the project? | 7 |
| 08 | Is there a formalized integration of the internal areas of the company: Engineering, Environment, Institutional Relationship and Social Responsibility? | 0 |
| 09 | Is there a formal process to monitor the release of Pre-Installation, Installation and Operation licenses from government agencies? | 0 |

## Energy Sector

The research was based on the Project management maturity study of an energy consulting company. Founded in 1993, with operations concentrated in the state of Minas Gerais, and which has big clients like Cemig, Vale, Alstom, ABB, and others.

As informed before, the project manager has been named by number in order to respect each one of them

### Project Manager 1

See Tables 82-90.

## Civil Construction Sector

The research was based on the Project management maturity study of an experienced company founded in 1995, with operations in Minas Gerais, Rio de Janeiro, São Paulo and Bahia. As informed before, the project manager has been named by number in order to respect each one of them.

### *Appendix*

## *Table 10. Project management*

| Id. | Question | Score |
|---|---|---|
| 01 | Is there any document to formalize the existence of the project in your area? | 7 |
| 02 | Does the project existence document adequately describe the project output? | 7 |
| 03 | Is there a document that can document the assumptions (prerequisites necessary for you to complete the work) and constraints (dates imposed by the client, limitation of funding, limitation of resources) of the project? | 7 |
| 04 | Is there a Project Analytical Framework (EAP) or WBS? | 7 |
| 05 | Is there any tool to formalize the client's acceptance? | 7 |
| 06 | Is there a baseline against which project progress, costs and deadlines are measured? | 7 |
| 07 | Is there an established process for project change control? | 7 |
| 08 | Is there a process of identifying predecessors and successors of project activities? | 7 |
| 09 | Is there a separation of project activities according to each schedule of the resource to be used (business days, holidays, etc.)? | 7 |
| 10 | Is there a process for identifying and prioritizing the critical path of the project? | 7 |
| 11 | Is there a baseline recording of the project at the end of the schedule development? | 7 |
| 12 | In developing the schedule, do all activities have resources? | 0 |
| 13 | Is there a formalization process of generating provisions (contingencies) for project costs? | 0 |
| 14 | Is the process of formalizing a matrix of responsibilities for the project? | 0 |
| 15 | Is there a formalized process of generating and transmitting project information during its execution (Communications Plan)? | 7 |
| 16 | Is there a mapping that details who needs the information, when it will be needed, how it should be defined, and by whom? | 7 |
| 17 | Is there a place with phones, addresses, e-mails and contacts from the entire team and involved? | 7 |
| 18 | Is there a formalized process for defining the schedule of project meetings? | 7 |
| 20 | Are the meetings pre-scheduled from the start and on the agenda of the entire team? | 7 |
| 21 | Does each team member know what information to bring to the follow-up meeting? | 7 |
| 22 | Is there a formal project budgeting process by grouping estimates of activities, resources, hiring, contingencies, etc.? | 0 |
| 23 | Is there a formalized risk identification process for the project? | 7 |
| 24 | Is there a qualitative and quantitative project risk assessment process? | 7 |
| 25 | Is there a project risk prioritization process? | 7 |
| 26 | Is there a process for planning the project's risk responses? | 7 |
| 27 | Is there a process to identify who is responsible for the project risk? | 7 |
| 28 | Is there a risk management and monitoring tool for the project? | 7 |
| 29 | Is there a process for identifying project product requirements? | 0 |
| 30 | Are the project product requirements translated into the work required to develop this product? | 0 |
| 31 | Is there a document that can formalize the work of the project (Scope Statement)? | 7 |
| 32 | Is there a formalized audit plan for quality requirements for the project (Quality Plan)? | 0 |
| 33 | Is there formalized planning for the provision and contracting of project resources (engineering, equipment, etc.)? | 7 |
| 34 | Is there a formalized contracting cycle for large project deliverables (Bid Cycle or Hiring Cycle)? | 7 |

*Table 11. Performance and results*

| Id. | Quesion | Score |
|---|---|---|
| 01 | Are deadlines met as planned? | 0 |
| 02 | Are the quality of the delivered projects according to the customer's requirements? | 0 |
| 03 | Are project costs met as planned? | 0 |
| 04 | Is there a monitoring of customer satisfaction with the project product? | 7 |
| 05 | Is there a process for evaluating the performance of the project manager in the conduct of the project? | 7 |
| 06 | Is there a project team evaluation process? | 7 |

*Table 12. Methods and practices*

| Id. | Question | Score |
|---|---|---|
| 01 | Before the start of the project is a feasibility study for the investment carried out? | 7 |
| 02 | Is there a practice of benchmarking projects in relation to projects developed by similar companies? | 7 |
| 03 | Is there practice of contracting consultancies to identify and evaluate project maturity for successful projects (IPA - Independent Project Analysis, etc.)? | 7 |
| 04 | Is there practice of hiring consultancies in the evaluation and optimization of budget? | 7 |
| 05 | Is there a formal process of closing contracts open during the project? | 0 |
| 06 | Is there a formal process of collecting the lessons learned during and at the end of the project? | 7 |
| 07 | Is there a formal process of cost control through tools like EVA - Value Added Estimation? | 0 |
| 08 | Is there a formal operational risk analysis and survey process (HAZOP - Hazard and Operability Studies)? | 7 |

## Project Manager 1

See Tables 91-99.

## Appendix

# *Table 13. Engineering and geology*

| Id. | Question | Score |
|---|---|---|
| 01 | Is there a formalized process that every project developed in the organization should go through the phases: conceptual, basic and detailed? | 7 |
| 02 | Is every project being run by the organization going through or going through the conceptual, basic, and detailed engineering phases? | 0 |
| 03 | Is there a formalized process for the drilling and auditing of mineral resources for engineering? | 7 |
| 04 | Is there a formal process that all conceptual engineering should begin with the first results of exploratory surveys? | 0 |
| 05 | Is there a formalization and submission of conceptual engineering information to the Environmental area for environmental licensing? | 0 |
| 06 | Is all Conceptual Engineering developed based on a mapping of biotopes (areas impacted by the environment) and surface? | 0 |
| 07 | Are all process tests (Technological Development) formally sent to Engineering to start Basic Engineering? | 0 |
| 08 | Is there a formalized process for reviewing engineering quality (Peer Review) before moving from one phase to another (eg: Conceptual Engineering for Basic Engineering)? | 0 |
| 09 | Is there a formalized engineering development culture in 3D technology? | 0 |
| 10 | Is there a process of evaluation, prior to the beginning of the Detailed Engineering, on the detailing of the project through the modularization of the construction? (Construction in Modules) | 7 |
| 11 | Is there the process of generating a Master Plan for the project after Conceptual Engineering? | 7 |
| 12 | If yes for answer 11, is there a formalized process of freezing this master plan and communicating it to all project stakeholders? | 0 |
| 13 | Is there a formalized process of generation of Descriptive Memorials, Spreadsheets, Data Sheet and Technical Specifications for the Supplies area? | 7 |
| 14 | Is there a formalized process of participation of the Supplies area during the evolution of the Basic Engineering for Detailed Engineering to determine the type of contracting of the Supply, Assembly and Implementation of the project? | 7 |
| 15 | Is there a formal process of realizing and studying tradeoffs for the lowest cost for utilities (water, gas, electricity, logistics, etc.)? | 7 |
| 16 | Is there a formal process of realization and study of opportunities (tradeoffs) regarding the location of the project (surface workers, communities, logistics, etc.)? | 0 |
| 17 | Is there a formal process for evaluating and validating each engineering phase (Conceptual, Basic, and Detailed) before moving from one to the other (Front End Loading Portions of Independent Project Analysis)? | 7 |
| 18 | At the end of the Project's Conceptual Engineering, is there a formal process to generate a Prefeasibility Study? | 7 |
| 19 | At the end of the Basic Engineering of the Project, is there a formal process to generate a Feasibility Study of the project? | 7 |
| 20 | During the execution of the Conceptual Engineering, is there a formal process of identification and first contacts with impacted landowners as well as local communities? | 7 |
| 21 | During the execution of Conceptual Engineering, is there a formal topographic survey of the region as well as the generation of orthophotos for the project's master plan? | 7 |
| 22 | After Conceptual Engineering, is there a formal process of generating investment estimates at a precision level of -15% to + 25%? | 0 |
| 23 | After the Basic Engineering, is there a formal process of generating the investment estimates at a precision level of -10% to + 10%? | 0 |
| 24 | Is there a formal closed-end planning process for the project after its useful life? | 0 |
| 25 | Is there a formal process for identifying VIPs - Value Improving Practices? | 7 |

*Table 14. Organizational structure*

| Id. | Question | Score |
|---|---|---|
| **01** | When a project starts, is there a project manager assigned by the organization to lead the project? | 0 |
| **02** | Is there an area focused on project planning? | 7 |
| **03** | Is there a formalized position of project manager within the organization? | 0 |
| **04** | Is there formalization of the Project Manager's power to similar functional positions (HR Manager, Operations Manager, etc.)? | 0 |
| **05** | Is there formalization of areas within Project manangement such as: Planning, PMO, Cost Control, etc.? | 0 |
| **06** | Are all internal areas of the company formally communicated from the beginning of the project and are their representatives formally assigned to the project? | 0 |

*Table 15. People management*

| Id. | Question | Score |
|---|---|---|
| **01** | Is there a formal process of developing the project manager's career and team? | 0 |
| **02** | Do follow-up meetings have a schedule and a fixed format? | 7 |
| **03** | Does each project member receive their work with instructions and in advance? | 7 |
| **04** | Is there a formal process of variable compensation for project team performance? | 7 |
| **05** | Is there a formal process for evaluating project team performance? | 7 |
| **06** | Is there a formal organization chart of the project team, delegations, etc.? | 0 |
| **07** | Are all the human resources required for the project formally sent to the HR area to arrange recruitment? | 0 |
| **08** | Are all interpersonal skills of the project team managed during the project? | 0 |
| **09** | Are all conflicts between the project team formally addressed? | 0 |

## Table 16. Project management office

| Id. | Question | Score |
|---|---|---|
| 01 | Is there a flow for the arrival in your area of the information about the existence of a project or a contract? | 7 |
| 02 | Is there a dedication of the professionals to the management of the projects? | 7 |
| 03 | Is there any tool for estimating effort (time) and project costs? | 7 |
| 04 | Is there any tool for estimating project resources? | 0 |
| 05 | Is there any tool to formalize the client's acceptance? | 7 |
| 06 | Is there an established process for formal acceptance of each deliverable of the project? | 0 |
| 07 | For the development of project documentation, is there a repository of past project templates for reuse in new projects? | 7 |
| 08 | Is there an area focused on the physical and financial control of the project? | 7 |
| 09 | Is there a visibility room, a mural or even an intranet site where everyone can access projects in a simple way? | 7 |
| 10 | In the case of geographically dispersed teams, are the procedures / technologies used to integrate them adequate? | 0 |
| 11 | Are there managerial reports with a previously defined format agreed upon with the client? | 7 |
| 12 | Is there a formalized process for the closing of a project / contract? | 0 |
| 13 | Is there a formal process for the provision of information in the closing phase (finalization of the work)? | 7 |
| 14 | There is an area focused on the control and support of the project (PMO or Project Office) | 7 |

## Table 17. Stakeholder management

| Id. | Question | Score |
|---|---|---|
| 01 | Is there a formalized process of mapping and managing the company's internal stakeholders? | 7 |
| 02 | Is there a formalized process for managing stakeholder expectations of the project during its execution? | 7 |
| 03 | Are formal interviews conducted to identify project stakeholders? | 0 |
| 04 | Are all potential impacts caused by each stakeholder formally assessed and managed during and after the project? | 0 |
| 05 | Is there a formal process of ranking and prioritizing the degree of influence of each stakeholder on the project? | 7 |
| 06 | Is there a formalized process of classification of stakeholder groups? | 7 |
| 07 | Is there a formal process of simulating the reactions of each stakeholder to the various situations that may occur during and after the project? | 0 |
| 08 | Is there a formalized process of gathering and filing the information of each stakeholder such as: name, address, telephone, etc.? | 0 |
| 09 | Is there a formalized Communication Plan capable of identifying what kind of information, when, how and for which stakeholder to send? | 0 |

*Table 18. Environment management*

| Id. | Question | Score |
| --- | --- | --- |
| 01 | Is there a formalized process of beginning environmental studies with the beginning of Conceptual Engineering? | 0 |
| 02 | Is there a formalized process for mapping biotopes (areas of environmental impact) to aid in the development of Conceptual Engineering? | 7 |
| 03 | Is there a formalization process of mapping the surface properties impacted by the project during the Environmental Licensing and Conceptual Engineering? | 0 |
| 04 | Is there a formal process of generating documentation for Environmental Impact Studies and Environmental Impact Report by the Engineering area? | 0 |
| 05 | Is there a formal process that synchronizes the development of Engineering and Environment activities? (Generation of Documentation for Environmental Licensing)? | 7 |
| 06 | Is there a formalized area for institutional relations with government agencies? | 7 |
| 07 | Is there a formalized area for relationship with the communities impacted by the project? | 7 |
| 08 | Is there a formalized integration of the internal areas of the company: Engineering, Environment, Institutional Relationship and Social Responsibility? | 7 |
| 09 | Is there a formal process to monitor the release of Pre-Installation, Installation and Operation licenses from government agencies? | 7 |

## Appendix

*Table 19. Project management*

| Id. | Question | Score |
|---|---|---|
| 01 | Is there any document to formalize the existence of the project in your area? | 7 |
| 02 | Does the project existence document adequately describe the project output? | 7 |
| 03 | Is there a document that can document the assumptions (prerequisites necessary for you to complete the work) and constraints (dates imposed by the client, limitation of funding, limitation of resources) of the project? | 0 |
| 04 | Is there a Project Analytical Framework (EAP) or WBS? | 7 |
| 05 | Is there any tool to formalize the client's acceptance? | 0 |
| 06 | Is there a baseline against which project progress, costs and deadlines are measured? | 7 |
| 07 | Is there an established process for project change control? | 0 |
| 08 | Is there a process of identifying predecessors and successors of project activities? | 7 |
| 09 | Is there a separation of project activities according to each schedule of the resource to be used (business days, holidays, etc.)? | 0 |
| 10 | Is there a process for identifying and prioritizing the critical path of the project? | 7 |
| 11 | Is there a baseline recording of the project at the end of the schedule development? | 7 |
| 12 | In developing the schedule, do all activities have resources? | 0 |
| 13 | Is there a formalization process of generating provisions (contingencies) for project costs? | 0 |
| 14 | Is the process of formalizing a matrix of responsibilities for the project? | 0 |
| 15 | Is there a formalized process of generating and transmitting project information during its execution (Communications Plan)? | 0 |
| 16 | Is there a mapping that details who needs the information, when it will be needed, how it should be defined, and by whom? | 7 |
| 17 | Is there a place with phones, addresses, e-mails and contacts from the entire team and involved? | 7 |
| 18 | Is there a formalized process for defining the schedule of project meetings? | 7 |
| 20 | Are the meetings pre-scheduled from the start and on the agenda of the entire team? | 7 |
| 21 | Does each team member know what information to bring to the follow-up meeting? | 7 |
| 22 | Is there a formal project budgeting process by grouping estimates of activities, resources, hiring, contingencies, etc.? | 0 |
| 23 | Is there a formalized risk identification process for the project? | 0 |
| 24 | Is there a qualitative and quantitative project risk assessment process? | 0 |
| 25 | Is there a project risk prioritization process? | 0 |
| 26 | Is there a process for planning the project's risk responses? | 0 |
| 27 | Is there a process to identify who is responsible for the project risk? | 0 |
| 28 | Is there a risk management and monitoring tool for the project? | 0 |
| 29 | Is there a process for identifying project product requirements? | 7 |
| 30 | Are the project product requirements translated into the work required to develop this product? | 7 |
| 31 | Is there a document that can formalize the work of the project (Scope Statement)? | 7 |
| 32 | Is there a formalized audit plan for quality requirements for the project (Quality Plan)? | 7 |
| 33 | Is there formalized planning for the provision and contracting of project resources (engineering, equipment, etc.)? | 0 |
| 34 | Is there a formalized contracting cycle for large project deliverables (Bid Cycle or Hiring Cycle)? | 0 |

*Table 20. Performance and results*

| Id. | Question | Score |
|---|---|---|
| 01 | Are deadlines met as planned? | 7 |
| 02 | Are the quality of the delivered projects according to the customer's requirements? | 7 |
| 03 | Are project costs met as planned? | 0 |
| 04 | Is there a monitoring of customer satisfaction with the project product? | 0 |
| 05 | Is there a process for evaluating the performance of the project manager in the conduct of the project? | 0 |
| 06 | Is there a project team evaluation process? | 7 |

*Table 21. Methods and practices*

| Id. | Question | Score |
|---|---|---|
| 01 | Before the start of the project is a feasibility study for the investment carried out? | 7 |
| 02 | Is there a practice of benchmarking projects in relation to projects developed by similar companies? | 7 |
| 03 | Is there practice of contracting consultancies to identify and evaluate project maturity for successful projects (IPA - Independent Project Analysis, etc.)? | 7 |
| 04 | Is there practice of hiring consultancies in the evaluation and optimization of budget? | 7 |
| 05 | Is there a formal process of closing contracts open during the project? | 7 |
| 06 | Is there a formal process of collecting the lessons learned during and at the end of the project? | 7 |
| 07 | Is there a formal process of cost control through tools like EVA - Value Added Estimation? | 0 |
| 08 | Is there a formal operational risk analysis and survey process (HAZOP - Hazard and Operability Studies)? | 7 |

## Appendix

# *Table 22. Engineering and geology*

| Id. | Question | Score |
|---|---|---|
| 01 | Is there a formalized process that every project developed in the organization should go through the phases: conceptual, basic and detailed? | 7 |
| 02 | Is every project being run by the organization going through or going through the conceptual, basic, and detailed engineering phases? | 0 |
| 03 | Is there a formalized process for the drilling and auditing of mineral resources for engineering? | 0 |
| 04 | Is there a formal process that all conceptual engineering should begin with the first results of exploratory surveys? | 0 |
| 05 | Is there a formalization and submission of conceptual engineering information to the Environmental area for environmental licensing? | 0 |
| 06 | Is all Conceptual Engineering developed based on a mapping of biotopes (areas impacted by the environment) and surface? | 0 |
| 07 | Are all process tests (Technological Development) formally sent to Engineering to start Basic Engineering? | 7 |
| 08 | Is there a formalized process for reviewing engineering quality (Peer Review) before moving from one phase to another (eg: Conceptual Engineering for Basic Engineering)? | 7 |
| 09 | Is there a formalized engineering development culture in 3D technology? | 0 |
| 10 | Is there a process of evaluation, prior to the beginning of the Detailed Engineering, on the detailing of the project through the modularization of the construction? (Construction in Modules) | 7 |
| 11 | Is there the process of generating a Master Plan for the project after Conceptual Engineering? | 7 |
| 12 | If yes for answer 11, is there a formalized process of freezing this master plan and communicating it to all project stakeholders? | 7 |
| 13 | Is there a formalized process of generation of Descriptive Memorials, Spreadsheets, Data Sheet and Technical Specifications for the Supplies area? | 7 |
| 14 | Is there a formalized process of participation of the Supplies area during the evolution of the Basic Engineering for Detailed Engineering to determine the type of contracting of the Supply, Assembly and Implementation of the project? | 0 |
| 15 | Is there a formal process of realizing and studying tradeoffs for the lowest cost for utilities (water, gas, electricity, logistics, etc.)? | 7 |
| 16 | Is there a formal process of realization and study of opportunities (tradeoffs) regarding the location of the project (surface workers, communities, logistics, etc.)? | 7 |
| 17 | Is there a formal process for evaluating and validating each engineering phase (Conceptual, Basic, and Detailed) before moving from one to the other (Front End Loading Portions of Independent Project Analysis)? | 7 |
| 18 | At the end of the Project's Conceptual Engineering, is there a formal process to generate a Prefeasibility Study? | 7 |
| 19 | At the end of the Basic Engineering of the Project, is there a formal process to generate a Feasibility Study of the project? | 7 |
| 20 | During the execution of the Conceptual Engineering, is there a formal process of identification and first contacts with impacted landowners as well as local communities? | 7 |
| 21 | During the execution of Conceptual Engineering, is there a formal topographic survey of the region as well as the generation of orthophotos for the project's master plan? | 7 |
| 22 | After Conceptual Engineering, is there a formal process of generating investment estimates at a precision level of -15% to + 25%? | 7 |
| 23 | After the Basic Engineering, is there a formal process of generating the investment estimates at a precision level of -10% to + 10%? | 7 |
| 24 | Is there a formal closed-end planning process for the project after its useful life? | 0 |
| 25 | Is there a formal process for identifying VIPs - Value Improving Practices? | 0 |

*Table 23. Organizational structure*

| Id. | Question | Score |
|---|---|---|
| **01** | When a project starts, is there a project manager assigned by the organization to lead the project? | 7 |
| **02** | Is there an area focused on project planning? | 7 |
| **03** | Is there a formalized position of project manager within the organization? | 7 |
| **04** | Is there formalization of the Project Manager's power to similar functional positions (HR Manager, Operations Manager, etc.)? | 7 |
| **05** | Is there formalization of areas within Project manangement such as: Planning, PMO, Cost Control, etc.? | 7 |
| **06** | Are all internal areas of the company formally communicated from the beginning of the project and are their representatives formally assigned to the project? | 7 |

*Table 24. People management*

| Id. | Question | Score |
|---|---|---|
| **01** | Is there a formal process of developing the project manager's career and team? | 0 |
| **02** | Do follow-up meetings have a schedule and a fixed format? | 0 |
| **03** | Does each project member receive their work with instructions and in advance? | 7 |
| **04** | Is there a formal process of variable compensation for project team performance? | 0 |
| **05** | Is there a formal process for evaluating project team performance? | 7 |
| **06** | Is there a formal organization chart of the project team, delegations, etc.? | 7 |
| **07** | Are all the human resources required for the project formally sent to the HR area to arrange recruitment? | 7 |
| **08** | Are all interpersonal skills of the project team managed during the project? | 7 |
| **09** | Are all conflicts between the project team formally addressed? | 7 |

*Table 25. Project manangement office*

| Id. | Question | Score |
|---|---|---|
| 01 | Is there a flow for the arrival in your area of the information about the existence of a project or a contract? | 7 |
| 02 | Is there a dedication of the professionals to the management of the projects? | 7 |
| 03 | Is there any tool for estimating effort (time) and project costs? | 7 |
| 04 | Is there any tool for estimating project resources? | 7 |
| 05 | Is there any tool to formalize the client's acceptance? | 0 |
| 06 | Is there an established process for formal acceptance of each deliverable of the project? | 0 |
| 07 | For the development of project documentation, is there a repository of past project templates for reuse in new projects? | 0 |
| 08 | Is there an area focused on the physical and financial control of the project? | 7 |
| 09 | Is there a visibility room, a mural or even an intranet site where everyone can access projects in a simple way? | 7 |
| 10 | In the case of geographically dispersed teams, are the procedures / technologies used to integrate them adequate? | 0 |
| 11 | Are there managerial reports with a previously defined format agreed upon with the client? | 7 |
| 12 | Is there a formalized process for the closing of a project / contract? | 7 |
| 13 | Is there a formal process for the provision of information in the closing phase (finalization of the work)? | 0 |
| 14 | There is an area focused on the control and support of the project (PMO or Project Office) | 7 |

*Table 26. Stakeholder management*

| Id. | Question | Score |
|---|---|---|
| 01 | Is there a formalized process of mapping and managing the company's internal stakeholders? | 0 |
| 02 | Is there a formalized process for managing stakeholder expectations of the project during its execution? | 7 |
| 03 | Are formal interviews conducted to identify project stakeholders? | 0 |
| 04 | Are all potential impacts caused by each stakeholder formally assessed and managed during and after the project? | 0 |
| 05 | Is there a formal process of ranking and prioritizing the degree of influence of each stakeholder on the project? | 0 |
| 06 | Is there a formalized process of classification of stakeholder groups? | 0 |
| 07 | Is there a formal process of simulating the reactions of each stakeholder to the various situations that may occur during and after the project? | 0 |
| 08 | Is there a formalized process of gathering and filing the information of each stakeholder such as: name, address, telephone, etc.? | 7 |
| 09 | Is there a formalized Communication Plan capable of identifying what kind of information, when, how and for which stakeholder to send? | 0 |

*Table 27. Environment management*

| Id. | Question | Score |
|---|---|---|
| 01 | Is there a formalized process of beginning environmental studies with the beginning of Conceptual Engineering? | 0 |
| 02 | Is there a formalized process for mapping biotopes (areas of environmental impact) to aid in the development of Conceptual Engineering? | 0 |
| 03 | Is there a formalization process of mapping the surface properties impacted by the project during the Environmental Licensing and Conceptual Engineering? | 7 |
| 04 | Is there a formal process of generating documentation for Environmental Impact Studies and Environmental Impact Report by the Engineering area? | 0 |
| 05 | Is there a formal process that synchronizes the development of Engineering and Environment activities? (Generation of Documentation for Environmental Licensing)? | 0 |
| 06 | Is there a formalized area for institutional relations with government agencies? | 7 |
| 07 | Is there a formalized area for relationship with the communities impacted by the project? | 7 |
| 08 | Is there a formalized integration of the internal areas of the company: Engineering, Environment, Institutional Relationship and Social Responsibility? | 7 |
| 09 | Is there a formal process to monitor the release of Pre-Installation, Installation and Operation licenses from government agencies? | 0 |

## *Appendix*

# *Table 28. Project management*

| Id. | Question | Score |
|---|---|---|
| 01 | Is there any document to formalize the existence of the project in your area? | 0 |
| 02 | Does the project existence document adequately describe the project output? | 0 |
| 03 | Is there a document that can document the assumptions (prerequisites necessary for you to complete the work) and constraints (dates imposed by the client, limitation of funding, limitation of resources) of the project? | 0 |
| 04 | Is there a Project Analytical Framework (EAP) or WBS? | 7 |
| 05 | Is there any tool to formalize the client's acceptance? | 7 |
| 06 | Is there a baseline against which project progress, costs and deadlines are measured? | 7 |
| 07 | Is there an established process for project change control? | 0 |
| 08 | Is there a process of identifying predecessors and successors of project activities? | 7 |
| 09 | Is there a separation of project activities according to each schedule of the resource to be used (business days, holidays, etc.)? | 0 |
| 10 | Is there a process for identifying and prioritizing the critical path of the project? | 7 |
| 11 | Is there a baseline recording of the project at the end of the schedule development? | 7 |
| 12 | In developing the schedule, do all activities have resources? | 0 |
| 13 | Is there a formalization process of generating provisions (contingencies) for project costs? | 0 |
| 14 | Is the process of formalizing a matrix of responsibilities for the project? | 0 |
| 15 | Is there a formalized process of generating and transmitting project information during its execution (Communications Plan)? | 0 |
| 16 | Is there a mapping that details who needs the information, when it will be needed, how it should be defined, and by whom? | 0 |
| 17 | Is there a place with phones, addresses, e-mails and contacts from the entire team and involved? | 0 |
| 18 | Is there a formalized process for defining the schedule of project meetings? | 7 |
| 20 | Are the meetings pre-scheduled from the start and on the agenda of the entire team? | 7 |
| 21 | Does each team member know what information to bring to the follow-up meeting? | 7 |
| 22 | Is there a formal project budgeting process by grouping estimates of activities, resources, hiring, contingencies, etc.? | 7 |
| 23 | Is there a formalized risk identification process for the project? | 7 |
| 24 | Is there a qualitative and quantitative project risk assessment process? | 7 |
| 25 | Is there a project risk prioritization process? | 7 |
| 26 | Is there a process for planning the project's risk responses? | 7 |
| 27 | Is there a process to identify who is responsible for the project risk? | 7 |
| 28 | Is there a risk management and monitoring tool for the project? | 7 |
| 29 | Is there a process for identifying project product requirements? | 0 |
| 30 | Are the project product requirements translated into the work required to develop this product? | 7 |
| 31 | Is there a document that can formalize the work of the project (Scope Statement)? | 0 |
| 32 | Is there a formalized audit plan for quality requirements for the project (Quality Plan)? | 0 |
| 33 | Is there formalized planning for the provision and contracting of project resources (engineering, equipment, etc.)? | 7 |
| 34 | Is there a formalized contracting cycle for large project deliverables (Bid Cycle or Hiring Cycle)? | 7 |

*Table 29. Performance and results*

| Id. | Quesion | Score |
|---|---|---|
| 01 | Are deadlines met as planned? | 0 |
| 02 | Are the quality of the delivered projects according to the customer's requirements? | 0 |
| 03 | Are project costs met as planned? | 0 |
| 04 | Is there a monitoring of customer satisfaction with the project product? | 0 |
| 05 | Is there a process for evaluating the performance of the project manager in the conduct of the project? | 7 |
| 06 | Is there a project team evaluation process? | 7 |

*Table 30. Methods and practices*

| Id. | Question | Score |
|---|---|---|
| 01 | Before the start of the project is a feasibility study for the investment carried out? | 7 |
| 02 | Is there a practice of benchmarking projects in relation to projects developed by similar companies? | 7 |
| 03 | Is there practice of contracting consultancies to identify and evaluate project maturity for successful projects (IPA - Independent Project Analysis, etc.)? | 7 |
| 04 | Is there practice of hiring consultancies in the evaluation and optimization of budget? | 7 |
| 05 | Is there a formal process of closing contracts open during the project? | 7 |
| 06 | Is there a formal process of collecting the lessons learned during and at the end of the project? | 0 |
| 07 | Is there a formal process of cost control through tools like EVA - Value Added Estimation? | 0 |
| 08 | Is there a formal operational risk analysis and survey process (HAZOP - Hazard and Operability Studies)? | 7 |

# Table 31. Engineering and geology

| Id. | Question | Score |
|---|---|---|
| 01 | Is there a formalized process that every project developed in the organization should go through the phases: conceptual, basic and detailed? | 0 |
| 02 | Is every project being run by the organization going through or going through the conceptual, basic, and detailed engineering phases? | 0 |
| 03 | Is there a formalized process for the drilling and auditing of mineral resources for engineering? | 7 |
| 04 | Is there a formal process that all conceptual engineering should begin with the first results of exploratory surveys? | 0 |
| 05 | Is there a formalization and submission of conceptual engineering information to the Environmental area for environmental licensing? | 7 |
| 06 | Is all Conceptual Engineering developed based on a mapping of biotopes (areas impacted by the environment) and surface? | 7 |
| 07 | Are all process tests (Technological Development) formally sent to Engineering to start Basic Engineering? | 7 |
| 08 | Is there a formalized process for reviewing engineering quality (Peer Review) before moving from one phase to another (eg: Conceptual Engineering for Basic Engineering)? | 0 |
| 09 | Is there a formalized engineering development culture in 3D technology? | 7 |
| 10 | Is there a process of evaluation, prior to the beginning of the Detailed Engineering, on the detailing of the project through the modularization of the construction? (Construction in Modules) | 7 |
| 11 | Is there the process of generating a Master Plan for the project after Conceptual Engineering? | 7 |
| 12 | If yes for answer 11, is there a formalized process of freezing this master plan and communicating it to all project stakeholders? | 0 |
| 13 | Is there a formalized process of generation of Descriptive Memorials, Spreadsheets, Data Sheet and Technical Specifications for the Supplies area? | 7 |
| 14 | Is there a formalized process of participation of the Supplies area during the evolution of the Basic Engineering for Detailed Engineering to determine the type of contracting of the Supply, Assembly and Implementation of the project? | 0 |
| 15 | Is there a formal process of realizing and studying tradeoffs for the lowest cost for utilities (water, gas, electricity, logistics, etc.)? | 7 |
| 16 | Is there a formal process of realization and study of opportunities (tradeoffs) regarding the location of the project (surface workers, communities, logistics, etc.)? | 7 |
| 17 | Is there a formal process for evaluating and validating each engineering phase (Conceptual, Basic, and Detailed) before moving from one to the other (Front End Loading Portions of Independent Project Analysis)? | 0 |
| 18 | At the end of the Project's Conceptual Engineering, is there a formal process to generate a Prefeasibility Study? | 0 |
| 19 | At the end of the Basic Engineering of the Project, is there a formal process to generate a Feasibility Study of the project? | 0 |
| 20 | During the execution of the Conceptual Engineering, is there a formal process of identification and first contacts with impacted landowners as well as local communities? | 7 |
| 21 | During the execution of Conceptual Engineering, is there a formal topographic survey of the region as well as the generation of orthophotos for the project's master plan? | 7 |
| 22 | After Conceptual Engineering, is there a formal process of generating investment estimates at a precision level of -15% to + 25%? | 0 |
| 23 | After the Basic Engineering, is there a formal process of generating the investment estimates at a precision level of -10% to + 10%? | 0 |
| 24 | Is there a formal closed-end planning process for the project after its useful life? | 0 |
| 25 | Is there a formal process for identifying VIPs - Value Improving Practices? | 7 |

*Table 32. Organizational structure*

| Id. | Question | Score |
|---|---|---|
| **01** | When a project starts, is there a project manager assigned by the organization to lead the project? | 7 |
| **02** | Is there an area focused on project planning? | 7 |
| **03** | Is there a formalized position of project manager within the organization? | 7 |
| **04** | Is there formalization of the Project Manager's power to similar functional positions (HR Manager, Operations Manager, etc.)? | 7 |
| **05** | Is there formalization of areas within Project manangement such as: Planning, PMO, Cost Control, etc.? | 7 |
| **06** | Are all internal areas of the company formally communicated from the beginning of the project and are their representatives formally assigned to the project? | 0 |

*Table 33. People management*

| Id. | Question | Score |
|---|---|---|
| **01** | Is there a formal process of developing the project manager's career and team? | 0 |
| **02** | Do follow-up meetings have a schedule and a fixed format? | 7 |
| **03** | Does each project member receive their work with instructions and in advance? | 0 |
| **04** | Is there a formal process of variable compensation for project team performance? | 7 |
| **05** | Is there a formal process for evaluating project team performance? | 7 |
| **06** | Is there a formal organization chart of the project team, delegations, etc.? | 7 |
| **07** | Are all the human resources required for the project formally sent to the HR area to arrange recruitment? | 7 |
| **08** | Are all interpersonal skills of the project team managed during the project? | 7 |
| **09** | Are all conflicts between the project team formally addressed? | 0 |

## *Table 34. Project management office*

| Id. | Question | Score |
|---|---|---|
| 01 | Is there a flow for the arrival in your area of the information about the existence of a project or a contract? | 0 |
| 02 | Is there a dedication of the professionals to the management of the projects? | 7 |
| 03 | Is there any tool for estimating effort (time) and project costs? | 0 |
| 04 | Is there any tool for estimating project resources? | 0 |
| 05 | Is there any tool to formalize the client's acceptance? | 7 |
| 06 | Is there an established process for formal acceptance of each deliverable of the project? | 7 |
| 07 | For the development of project documentation, is there a repository of past project templates for reuse in new projects? | 7 |
| 08 | Is there an area focused on the physical and financial control of the project? | 7 |
| 09 | Is there a visibility room, a mural or even an intranet site where everyone can access projects in a simple way? | 0 |
| 10 | In the case of geographically dispersed teams, are the procedures / technologies used to integrate them adequate? | |
| 11 | Are there managerial reports with a previously defined format agreed upon with the client? | 0 |
| 12 | Is there a formalized process for the closing of a project / contract? | 7 |
| 13 | Is there a formal process for the provision of information in the closing phase (finalization of the work)? | 0 |
| 14 | There is an area focused on the control and support of the project (PMO or Project Office) | 7 |

## *Table 35. Stakeholder management*

| Id. | Question | Score |
|---|---|---|
| 01 | Is there a formalized process of mapping and managing the company's internal stakeholders? | 7 |
| 02 | Is there a formalized process for managing stakeholder expectations of the project during its execution? | 7 |
| 03 | Are formal interviews conducted to identify project stakeholders? | 0 |
| 04 | Are all potential impacts caused by each stakeholder formally assessed and managed during and after the project? | 0 |
| 05 | Is there a formal process of ranking and prioritizing the degree of influence of each stakeholder on the project? | 7 |
| 06 | Is there a formalized process of classification of stakeholder groups? | 7 |
| 07 | Is there a formal process of simulating the reactions of each stakeholder to the various situations that may occur during and after the project? | 0 |
| 08 | Is there a formalized process of gathering and filing the information of each stakeholder such as: name, address, telephone, etc.? | 0 |
| 09 | Is there a formalized Communication Plan capable of identifying what kind of information, when, how and for which stakeholder to send? | 0 |

*Table 36. Environment management*

| Id. | Question | Score |
|---|---|---|
| 01 | Is there a formalized process of beginning environmental studies with the beginning of Conceptual Engineering? | 0 |
| 02 | Is there a formalized process for mapping biotopes (areas of environmental impact) to aid in the development of Conceptual Engineering? | 0 |
| 03 | Is there a formalization process of mapping the surface properties impacted by the project during the Environmental Licensing and Conceptual Engineering? | 0 |
| 04 | Is there a formal process of generating documentation for Environmental Impact Studies and Environmental Impact Report by the Engineering area? | 7 |
| 05 | Is there a formal process that synchronizes the development of Engineering and Environment activities? (Generation of Documentation for Environmental Licensing)? | 7 |
| 06 | Is there a formalized area for institutional relations with government agencies? | 7 |
| 07 | Is there a formalized area for relationship with the communities impacted by the project? | 7 |
| 08 | Is there a formalized integration of the internal areas of the company: Engineering, Environment, Institutional Relationship and Social Responsibility? | 0 |
| 09 | Is there a formal process to monitor the release of Pre-Installation, Installation and Operation licenses from government agencies? | 7 |

## Appendix

# *Table 37. Project management*

| Id. | Question | Score |
|---|---|---|
| 01 | Is there any document to formalize the existence of the project in your area? | 7 |
| 02 | Does the project existence document adequately describe the project output? | 0 |
| 03 | Is there a document that can document the assumptions (prerequisites necessary for you to complete the work) and constraints (dates imposed by the client, limitation of funding, limitation of resources) of the project? | 0 |
| 04 | Is there a Project Analytical Framework (EAP) or WBS? | 7 |
| 05 | Is there any tool to formalize the client's acceptance? | 7 |
| 06 | Is there a baseline against which project progress, costs and deadlines are measured? | 0 |
| 07 | Is there an established process for project change control? | 0 |
| 08 | Is there a process of identifying predecessors and successors of project activities? | 0 |
| 09 | Is there a separation of project activities according to each schedule of the resource to be used (business days, holidays, etc.)? | 0 |
| 10 | Is there a process for identifying and prioritizing the critical path of the project? | 7 |
| 11 | Is there a baseline recording of the project at the end of the schedule development? | 0 |
| 12 | In developing the schedule, do all activities have resources? | 0 |
| 13 | Is there a formalization process of generating provisions (contingencies) for project costs? | 0 |
| 14 | Is the process of formalizing a matrix of responsibilities for the project? | 7 |
| 15 | Is there a formalized process of generating and transmitting project information during its execution (Communications Plan)? | 7 |
| 16 | Is there a mapping that details who needs the information, when it will be needed, how it should be defined, and by whom? | 7 |
| 17 | Is there a place with phones, addresses, e-mails and contacts from the entire team and involved? | 0 |
| 18 | Is there a formalized process for defining the schedule of project meetings? | 0 |
| 20 | Are the meetings pre-scheduled from the start and on the agenda of the entire team? | 0 |
| 21 | Does each team member know what information to bring to the follow-up meeting? | 0 |
| 22 | Is there a formal project budgeting process by grouping estimates of activities, resources, hiring, contingencies, etc.? | 0 |
| 23 | Is there a formalized risk identification process for the project? | 7 |
| 24 | Is there a qualitative and quantitative project risk assessment process? | 7 |
| 25 | Is there a project risk prioritization process? | 7 |
| 26 | Is there a process for planning the project's risk responses? | 7 |
| 27 | Is there a process to identify who is responsible for the project risk? | 7 |
| 28 | Is there a risk management and monitoring tool for the project? | 0 |
| 29 | Is there a process for identifying project product requirements? | 0 |
| 30 | Are the project product requirements translated into the work required to develop this product? | 0 |
| 31 | Is there a document that can formalize the work of the project (Scope Statement)? | 0 |
| 32 | Is there a formalized audit plan for quality requirements for the project (Quality Plan)? | 0 |
| 33 | Is there formalized planning for the provision and contracting of project resources (engineering, equipment, etc.)? | 7 |
| 34 | Is there a formalized contracting cycle for large project deliverables (Bid Cycle or Hiring Cycle)? | 0 |

*Table 38. Performance and results*

| Id. | Question | Score |
|---|---|---|
| **01** | Are deadlines met as planned? | 0 |
| **02** | Are the quality of the delivered projects according to the customer's requirements? | 0 |
| **03** | Are project costs met as planned? | 0 |
| **04** | Is there a monitoring of customer satisfaction with the project product? | 0 |
| **05** | Is there a process for evaluating the performance of the project manager in the conduct of the project? | 0 |
| **06** | Is there a project team evaluation process? | 0 |

*Table 39. Methods and practices*

| Id. | Question | Score |
|---|---|---|
| **01** | Before the start of the project is a feasibility study for the investment carried out? | 7 |
| **02** | Is there a practice of benchmarking projects in relation to projects developed by similar companies? | 0 |
| **03** | Is there practice of contracting consultancies to identify and evaluate project maturity for successful projects (IPA - Independent Project Analysis, etc.)? | 7 |
| **04** | Is there practice of hiring consultancies in the evaluation and optimization of budget? | 7 |
| **05** | Is there a formal process of closing contracts open during the project? | 7 |
| **06** | Is there a formal process of collecting the lessons learned during and at the end of the project? | 0 |
| **07** | Is there a formal process of cost control through tools like EVA - Value Added Estimation? | 0 |
| **08** | Is there a formal operational risk analysis and survey process (HAZOP - Hazard and Operability Studies)? | 0 |

# Table 40. Engineering and geology

| Id. | Question | Score |
|---|---|---|
| 01 | Is there a formalized process that every project developed in the organization should go through the phases: conceptual, basic and detailed? | 0 |
| 02 | Is every project being run by the organization going through or going through the conceptual, basic, and detailed engineering phases? | 0 |
| 03 | Is there a formalized process for the drilling and auditing of mineral resources for engineering? | 0 |
| 04 | Is there a formal process that all conceptual engineering should begin with the first results of exploratory surveys? | 0 |
| 05 | Is there a formalization and submission of conceptual engineering information to the Environmental area for environmental licensing? | 0 |
| 06 | Is all Conceptual Engineering developed based on a mapping of biotopes (areas impacted by the environment) and surface? | 0 |
| 07 | Are all process tests (Technological Development) formally sent to Engineering to start Basic Engineering? | 0 |
| 08 | Is there a formalized process for reviewing engineering quality (Peer Review) before moving from one phase to another (eg: Conceptual Engineering for Basic Engineering)? | 0 |
| 09 | Is there a formalized engineering development culture in 3D technology? | 0 |
| 10 | Is there a process of evaluation, prior to the beginning of the Detailed Engineering, on the detailing of the project through the modularization of the construction? (Construction in Modules) | 7 |
| 11 | Is there the process of generating a Master Plan for the project after Conceptual Engineering? | 0 |
| 12 | If yes for answer 11, is there a formalized process of freezing this master plan and communicating it to all project stakeholders? | 0 |
| 13 | Is there a formalized process of generation of Descriptive Memorials, Spreadsheets, Data Sheet and Technical Specifications for the Supplies area? | 7 |
| 14 | Is there a formalized process of participation of the Supplies area during the evolution of the Basic Engineering for Detailed Engineering to determine the type of contracting of the Supply, Assembly and Implementation of the project? | 0 |
| 15 | Is there a formal process of realizing and studying tradeoffs for the lowest cost for utilities (water, gas, electricity, logistics, etc.)? | 0 |
| 16 | Is there a formal process of realization and study of opportunities (tradeoffs) regarding the location of the project (surface workers, communities, logistics, etc.)? | 0 |
| 17 | Is there a formal process for evaluating and validating each engineering phase (Conceptual, Basic, and Detailed) before moving from one to the other (Front End Loading Portions of Independent Project Analysis)? | 0 |
| 18 | At the end of the Project's Conceptual Engineering, is there a formal process to generate a Prefeasibility Study? | 0 |
| 19 | At the end of the Basic Engineering of the Project, is there a formal process to generate a Feasibility Study of the project? | 0 |
| 20 | During the execution of the Conceptual Engineering, is there a formal process of identification and first contacts with impacted landowners as well as local communities? | 0 |
| 21 | During the execution of Conceptual Engineering, is there a formal topographic survey of the region as well as the generation of orthophotos for the project's master plan? | 7 |
| 22 | After Conceptual Engineering, is there a formal process of generating investment estimates at a precision level of -15% to + 25%? | 0 |
| 23 | After the Basic Engineering, is there a formal process of generating the investment estimates at a precision level of -10% to + 10%? | 0 |
| 24 | Is there a formal closed-end planning process for the project after its useful life? | 0 |
| 25 | Is there a formal process for identifying VIPs - Value Improving Practices? | 0 |

*Table 41. Organizational structure*

| Id. | Question | Score |
|---|---|---|
| **01** | When a project starts, is there a project manager assigned by the organization to lead the project? | 0 |
| **02** | Is there an area focused on project planning? | 7 |
| **03** | Is there a formalized position of project manager within the organization? | 7 |
| **04** | Is there formalization of the Project Manager's power to similar functional positions (HR Manager, Operations Manager, etc.)? | 0 |
| **05** | Is there formalization of areas within Project management such as: Planning, PMO, Cost Control, etc.? | 7 |
| **06** | Are all internal areas of the company formally communicated from the beginning of the project and are their representatives formally assigned to the project? | 0 |

*Table 42. People management*

| Id. | Question | Score |
|---|---|---|
| **01** | Is there a formal process of developing the project manager's career and team? | 0 |
| **02** | Do follow-up meetings have a schedule and a fixed format? | 0 |
| **03** | Does each project member receive their work with instructions and in advance? | 0 |
| **04** | Is there a formal process of variable compensation for project team performance? | 0 |
| **05** | Is there a formal process for evaluating project team performance? | 7 |
| **06** | Is there a formal organization chart of the project team, delegations, etc.? | 7 |
| **07** | Are all the human resources required for the project formally sent to the HR area to arrange recruitment? | 7 |
| **08** | Are all interpersonal skills of the project team managed during the project? | 0 |
| **09** | Are all conflicts between the project team formally addressed? | 0 |

Appendix

*Table 43. Project management office*

| Id. | Question | Score |
|---|---|---|
| 01 | Is there a flow for the arrival in your area of the information about the existence of a project or a contract? | 0 |
| 02 | Is there a dedication of the professionals to the management of the projects? | 7 |
| 03 | Is there any tool for estimating effort (time) and project costs? | 7 |
| 04 | Is there any tool for estimating project resources? | 0 |
| 05 | Is there any tool to formalize the client's acceptance? | 7 |
| 06 | Is there an established process for formal acceptance of each deliverable of the project? | 0 |
| 07 | For the development of project documentation, is there a repository of past project templates for reuse in new projects? | 7 |
| 08 | Is there an area focused on the physical and financial control of the project? | 0 |
| 09 | Is there a visibility room, a mural or even an intranet site where everyone can access projects in a simple way? | 7 |
| 10 | In the case of geographically dispersed teams, are the procedures / technologies used to integrate them adequate? | 0 |
| 11 | Are there managerial reports with a previously defined format agreed upon with the client? | 7 |
| 12 | Is there a formalized process for the closing of a project / contract? | 7 |
| 13 | Is there a formal process for the provision of information in the closing phase (finalization of the work)? | 0 |
| 14 | There is an area focused on the control and support of the project (PMO or Project Office) | 7 |

*Table 44. Stakeholder management*

| Id. | Question | Score |
|---|---|---|
| 01 | Is there a formalized process of mapping and managing the company's internal stakeholders? | 7 |
| 02 | Is there a formalized process for managing stakeholder expectations of the project during its execution? | 0 |
| 03 | Are formal interviews conducted to identify project stakeholders? | 0 |
| 04 | Are all potential impacts caused by each stakeholder formally assessed and managed during and after the project? | 7 |
| 05 | Is there a formal process of ranking and prioritizing the degree of influence of each stakeholder on the project? | 0 |
| 06 | Is there a formalized process of classification of stakeholder groups? | 0 |
| 07 | Is there a formal process of simulating the reactions of each stakeholder to the various situations that may occur during and after the project? | 7 |
| 08 | Is there a formalized process of gathering and filing the information of each stakeholder such as: name, address, telephone, etc.? | 0 |
| 09 | Is there a formalized Communication Plan capable of identifying what kind of information, when, how and for which stakeholder to send? | 0 |

*Table 45. Environment management*

| Id. | Question | Score |
|---|---|---|
| 01 | Is there a formalized process of beginning environmental studies with the beginning of Conceptual Engineering? | 0 |
| 02 | Is there a formalized process for mapping biotopes (areas of environmental impact) to aid in the development of Conceptual Engineering? | 7 |
| 03 | Is there a formalization process of mapping the surface properties impacted by the project during the Environmental Licensing and Conceptual Engineering? | 0 |
| 04 | Is there a formal process of generating documentation for Environmental Impact Studies and Environmental Impact Report by the Engineering area? | 0 |
| 05 | Is there a formal process that synchronizes the development of Engineering and Environment activities? (Generation of Documentation for Environmental Licensing)? | 0 |
| 06 | Is there a formalized area for institutional relations with government agencies? | 0 |
| 07 | Is there a formalized area for relationship with the communities impacted by the project? | 7 |
| 08 | Is there a formalized integration of the internal areas of the company: Engineering, Environment, Institutional Relationship and Social Responsibility? | 0 |
| 09 | Is there a formal process to monitor the release of Pre-Installation, Installation and Operation licenses from government agencies? | 0 |

# Table 46. Project management

| Id. | Question | Score |
|---|---|---|
| 01 | Is there any document to formalize the existence of the project in your area? | 0 |
| 02 | Does the project existence document adequately describe the project output? | 0 |
| 03 | Is there a document that can document the assumptions (prerequisites necessary for you to complete the work) and constraints (dates imposed by the client, limitation of funding, limitation of resources) of the project? | 0 |
| 04 | Is there a Project Analytical Framework (EAP) or WBS? | 0 |
| 05 | Is there any tool to formalize the client's acceptance? | 0 |
| 06 | Is there a baseline against which project progress, costs and deadlines are measured? | 0 |
| 07 | Is there an established process for project change control? | 0 |
| 08 | Is there a process of identifying predecessors and successors of project activities? | 0 |
| 09 | Is there a separation of project activities according to each schedule of the resource to be used (business days, holidays, etc.)? | 0 |
| 10 | Is there a process for identifying and prioritizing the critical path of the project? | 0 |
| 11 | Is there a baseline recording of the project at the end of the schedule development? | 0 |
| 12 | In developing the schedule, do all activities have resources? | 0 |
| 13 | Is there a formalization process of generating provisions (contingencies) for project costs? | 0 |
| 14 | Is the process of formalizing a matrix of responsibilities for the project? | 0 |
| 15 | Is there a formalized process of generating and transmitting project information during its execution (Communications Plan)? | 0 |
| 16 | Is there a mapping that details who needs the information, when it will be needed, how it should be defined, and by whom? | 0 |
| 17 | Is there a place with phones, addresses, e-mails and contacts from the entire team and involved? | 0 |
| 18 | Is there a formalized process for defining the schedule of project meetings? | 0 |
| 20 | Are the meetings pre-scheduled from the start and on the agenda of the entire team? | 7 |
| 21 | Does each team member know what information to bring to the follow-up meeting? | 0 |
| 22 | Is there a formal project budgeting process by grouping estimates of activities, resources, hiring, contingencies, etc.? | 0 |
| 23 | Is there a formalized risk identification process for the project? | 0 |
| 24 | Is there a qualitative and quantitative project risk assessment process? | 0 |
| 25 | Is there a project risk prioritization process? | 0 |
| 26 | Is there a process for planning the project's risk responses? | 0 |
| 27 | Is there a process to identify who is responsible for the project risk? | 0 |
| 28 | Is there a risk management and monitoring tool for the project? | 0 |
| 29 | Is there a process for identifying project product requirements? | 0 |
| 30 | Are the project product requirements translated into the work required to develop this product? | 0 |
| 31 | Is there a document that can formalize the work of the project (Scope Statement)? | 0 |
| 32 | Is there a formalized audit plan for quality requirements for the project (Quality Plan)? | 0 |
| 33 | Is there formalized planning for the provision and contracting of project resources (engineering, equipment, etc.)? | 0 |
| 34 | Is there a formalized contracting cycle for large project deliverables (Bid Cycle or Hiring Cycle)? | 0 |

*Table 47. Performance and results*

| Id. | Question | Score |
|---|---|---|
| 01 | Are deadlines met as planned? | 0 |
| 02 | Are the quality of the delivered projects according to the customer's requirements? | 0 |
| 03 | Are project costs met as planned? | 0 |
| 04 | Is there a monitoring of customer satisfaction with the project product? | 0 |
| 05 | Is there a process for evaluating the performance of the project manager in the conduct of the project? | 0 |
| 06 | Is there a project team evaluation process? | 0 |

*Table 48. Methods and practices*

| Id. | Question | Score |
|---|---|---|
| 01 | Before the start of the project is a feasibility study for the investment carried out? | 0 |
| 02 | Is there a practice of benchmarking projects in relation to projects developed by similar companies? | 0 |
| 03 | Is there practice of contracting consultancies to identify and evaluate project maturity for successful projects (IPA - Independent Project Analysis, etc.)? | 0 |
| 04 | Is there practice of hiring consultancies in the evaluation and optimization of budget? | 0 |
| 05 | Is there a formal process of closing contracts open during the project? | 0 |
| 06 | Is there a formal process of collecting the lessons learned during and at the end of the project? | 0 |
| 07 | Is there a formal process of cost control through tools like EVA - Value Added Estimation? | 0 |
| 08 | Is there a formal operational risk analysis and survey process (HAZOP - Hazard and Operability Studies)? | 0 |

## Appendix

## *Table 49. Engineering and geology*

| Id. | Question | Score |
|---|---|---|
| 01 | Is there a formalized process that every project developed in the organization should go through the phases: conceptual, basic and detailed? | 0 |
| 02 | Is every project being run by the organization going through or going through the conceptual, basic, and detailed engineering phases? | 0 |
| 03 | Is there a formalized process for the drilling and auditing of mineral resources for engineering? | 0 |
| 04 | Is there a formal process that all conceptual engineering should begin with the first results of exploratory surveys? | 0 |
| 05 | Is there a formalization and submission of conceptual engineering information to the Environmental area for environmental licensing? | 0 |
| 06 | Is all Conceptual Engineering developed based on a mapping of biotopes (areas impacted by the environment) and surface? | 0 |
| 07 | Are all process tests (Technological Development) formally sent to Engineering to start Basic Engineering? | 0 |
| 08 | Is there a formalized process for reviewing engineering quality (Peer Review) before moving from one phase to another (eg: Conceptual Engineering for Basic Engineering)? | 0 |
| 09 | Is there a formalized engineering development culture in 3D technology? | 0 |
| 10 | Is there a process of evaluation, prior to the beginning of the Detailed Engineering, on the detailing of the project through the modularization of the construction? (Construction in Modules) | 0 |
| 11 | Is there the process of generating a Master Plan for the project after Conceptual Engineering? | 0 |
| 12 | If yes for answer 11, is there a formalized process of freezing this master plan and communicating it to all project stakeholders? | 0 |
| 13 | Is there a formalized process of generation of Descriptive Memorials, Spreadsheets, Data Sheet and Technical Specifications for the Supplies area? | 7 |
| 14 | Is there a formalized process of participation of the Supplies area during the evolution of the Basic Engineering for Detailed Engineering to determine the type of contracting of the Supply, Assembly and Implementation of the project? | 0 |
| 15 | Is there a formal process of realizing and studying tradeoffs for the lowest cost for utilities (water, gas, electricity, logistics, etc.)? | 0 |
| 16 | Is there a formal process of realization and study of opportunities (tradeoffs) regarding the location of the project (surface workers, communities, logistics, etc.)? | 0 |
| 17 | Is there a formal process for evaluating and validating each engineering phase (Conceptual, Basic, and Detailed) before moving from one to the other (Front End Loading Portions of Independent Project Analysis)? | 0 |
| 18 | At the end of the Project's Conceptual Engineering, is there a formal process to generate a Prefeasibility Study? | 0 |
| 19 | At the end of the Basic Engineering of the Project, is there a formal process to generate a Feasibility Study of the project? | 0 |
| 20 | During the execution of the Conceptual Engineering, is there a formal process of identification and first contacts with impacted landowners as well as local communities? | 0 |
| 21 | During the execution of Conceptual Engineering, is there a formal topographic survey of the region as well as the generation of orthophotos for the project's master plan? | 7 |
| 22 | After Conceptual Engineering, is there a formal process of generating investment estimates at a precision level of -15% to + 25%? | 0 |
| 23 | After the Basic Engineering, is there a formal process of generating the investment estimates at a precision level of -10% to + 10%? | 0 |
| 24 | Is there a formal closed-end planning process for the project after its useful life? | 0 |
| 25 | Is there a formal process for identifying VIPs - Value Improving Practices? | 0 |

*Table 50. Organizational structure*

| Id. | Question | Score |
|---|---|---|
| 01 | When a project starts, is there a project manager assigned by the organization to lead the project? | 0 |
| 02 | Is there an area focused on project planning? | 0 |
| 03 | Is there a formalized position of project manager within the organization? | 0 |
| 04 | Is there formalization of the Project Manager's power to similar functional positions (HR Manager, Operations Manager, etc.)? | 0 |
| 05 | Is there formalization of areas within Project manangement such as: Planning, PMO, Cost Control, etc.? | 0 |
| 06 | Are all internal areas of the company formally communicated from the beginning of the project and are their representatives formally assigned to the project? | 0 |

*Table 51. People management*

| Id. | Question | Score |
|---|---|---|
| 01 | Is there a formal process of developing the project manager's career and team? | 0 |
| 02 | Do follow-up meetings have a schedule and a fixed format? | 0 |
| 03 | Does each project member receive their work with instructions and in advance? | 0 |
| 04 | Is there a formal process of variable compensation for project team performance? | 0 |
| 05 | Is there a formal process for evaluating project team performance? | 7 |
| 06 | Is there a formal organization chart of the project team, delegations, etc.? | 0 |
| 07 | Are all the human resources required for the project formally sent to the HR area to arrange recruitment? | 0 |
| 08 | Are all interpersonal skills of the project team managed during the project? | 0 |
| 09 | Are all conflicts between the project team formally addressed? | 7 |

*Table 52. Project management office*

| Id. | Question | Score |
|---|---|---|
| 01 | Is there a flow for the arrival in your area of the information about the existence of a project or a contract? | 7 |
| 02 | Is there a dedication of the professionals to the management of the projects? | 7 |
| 03 | Is there any tool for estimating effort (time) and project costs? | 0 |
| 04 | Is there any tool for estimating project resources? | 0 |
| 05 | Is there any tool to formalize the client's acceptance? | 7 |
| 06 | Is there an established process for formal acceptance of each deliverable of the project? | 0 |
| 07 | For the development of project documentation, is there a repository of past project templates for reuse in new projects? | 0 |
| 08 | Is there an area focused on the physical and financial control of the project? | 7 |
| 09 | Is there a visibility room, a mural or even an intranet site where everyone can access projects in a simple way? | 0 |
| 10 | In the case of geographically dispersed teams, are the procedures / technologies used to integrate them adequate? | 0 |
| 11 | Are there managerial reports with a previously defined format agreed upon with the client? | 7 |
| 12 | Is there a formalized process for the closing of a project / contract? | 0 |
| 13 | Is there a formal process for the provision of information in the closing phase (finalization of the work)? | 0 |
| 14 | There is an area focused on the control and support of the project (PMO or Project Office) | 7 |

*Table 53. Stakeholder management*

| Id. | Question | Score |
|---|---|---|
| 01 | Is there a formalized process of mapping and managing the company's internal stakeholders? | 0 |
| 02 | Is there a formalized process for managing stakeholder expectations of the project during its execution? | 0 |
| 03 | Are formal interviews conducted to identify project stakeholders? | 0 |
| 04 | Are all potential impacts caused by each stakeholder formally assessed and managed during and after the project? | 0 |
| 05 | Is there a formal process of ranking and prioritizing the degree of influence of each stakeholder on the project? | 0 |
| 06 | Is there a formalized process of classification of stakeholder groups? | 0 |
| 07 | Is there a formal process of simulating the reactions of each stakeholder to the various situations that may occur during and after the project? | 0 |
| 08 | Is there a formalized process of gathering and filing the information of each stakeholder such as: name, address, telephone, etc.? | 0 |
| 09 | Is there a formalized Communication Plan capable of identifying what kind of information, when, how and for which stakeholder to send? | 0 |

*Table 54. Environment management*

| Id. | Question | Score |
|---|---|---|
| **01** | Is there a formalized process of beginning environmental studies with the beginning of Conceptual Engineering? | 0 |
| **02** | Is there a formalized process for mapping biotopes (areas of environmental impact) to aid in the development of Conceptual Engineering? | 0 |
| **03** | Is there a formalization process of mapping the surface properties impacted by the project during the Environmental Licensing and Conceptual Engineering? | 0 |
| **04** | Is there a formal process of generating documentation for Environmental Impact Studies and Environmental Impact Report by the Engineering area? | 0 |
| **05** | Is there a formal process that synchronizes the development of Engineering and Environment activities? (Generation of Documentation for Environmental Licensing)? | 0 |
| **06** | Is there a formalized area for institutional relations with government agencies? | 7 |
| **07** | Is there a formalized area for relationship with the communities impacted by the project? | 7 |
| **08** | Is there a formalized integration of the internal areas of the company: Engineering, Environment, Institutional Relationship and Social Responsibility? | 0 |
| **09** | Is there a formal process to monitor the release of Pre-Installation, Installation and Operation licenses from government agencies? | 0 |

*Table 55. Project management*

| Id. | Question | Score |
|---|---|---|
| 01 | Is there any document to formalize the existence of the project in your area? | 7 |
| 02 | Does the project existence document adequately describe the project output? | 7 |
| 03 | Is there a document that can document the assumptions (prerequisites necessary for you to complete the work) and constraints (dates imposed by the client, limitation of funding, limitation of resources) of the project? | 0 |
| 04 | Is there a Project Analytical Framework (EAP) or WBS? | 7 |
| 05 | Is there any tool to formalize the client's acceptance? | 0 |
| 06 | Is there a baseline against which project progress, costs and deadlines are measured? | 7 |
| 07 | Is there an established process for project change control? | 0 |
| 08 | Is there a process of identifying predecessors and successors of project activities? | 0 |
| 09 | Is there a separation of project activities according to each schedule of the resource to be used (business days, holidays, etc.)? | 7 |
| 10 | Is there a process for identifying and prioritizing the critical path of the project? | 0 |
| 11 | Is there a baseline recording of the project at the end of the schedule development? | 0 |
| 12 | In developing the schedule, do all activities have resources? | 0 |
| 13 | Is there a formalization process of generating provisions (contingencies) for project costs? | 0 |
| 14 | Is the process of formalizing a matrix of responsibilities for the project? | 7 |
| 15 | Is there a formalized process of generating and transmitting project information during its execution (Communications Plan)? | 0 |
| 16 | Is there a mapping that details who needs the information, when it will be needed, how it should be defined, and by whom? | 0 |
| 17 | Is there a place with phones, addresses, e-mails and contacts from the entire team and involved? | 0 |
| 18 | Is there a formalized process for defining the schedule of project meetings? | 7 |
| 20 | Are the meetings pre-scheduled from the start and on the agenda of the entire team? | 7 |
| 21 | Does each team member know what information to bring to the follow-up meeting? | 7 |
| 22 | Is there a formal project budgeting process by grouping estimates of activities, resources, hiring, contingencies, etc.? | 0 |
| 23 | Is there a formalized risk identification process for the project? | 7 |
| 24 | Is there a qualitative and quantitative project risk assessment process? | 7 |
| 25 | Is there a project risk prioritization process? | 7 |
| 26 | Is there a process for planning the project's risk responses? | 7 |
| 27 | Is there a process to identify who is responsible for the project risk? | 7 |
| 28 | Is there a risk management and monitoring tool for the project? | 7 |
| 29 | Is there a process for identifying project product requirements? | 7 |
| 30 | Are the project product requirements translated into the work required to develop this product? | 0 |
| 31 | Is there a document that can formalize the work of the project (Scope Statement)? | 7 |
| 32 | Is there a formalized audit plan for quality requirements for the project (Quality Plan)? | 0 |
| 33 | Is there formalized planning for the provision and contracting of project resources (engineering, equipment, etc.)? | 0 |
| 34 | Is there a formalized contracting cycle for large project deliverables (Bid Cycle or Hiring Cycle)? | 0 |

*Table 56. Performance and results*

| Id. | Quesion | Score |
|---|---|---|
| **01** | Are deadlines met as planned? | 0 |
| **02** | Are the quality of the delivered projects according to the customer's requirements? | 7 |
| **03** | Are project costs met as planned? | 7 |
| **04** | Is there a monitoring of customer satisfaction with the project product? | 0 |
| **05** | Is there a process for evaluating the performance of the project manager in the conduct of the project? | 0 |
| **06** | Is there a project team evaluation process? | 7 |

*Table 57. Methods and practices*

| Id. | Question | Score |
|---|---|---|
| **01** | Before the start of the project is a feasibility study for the investment carried out? | 7 |
| **02** | Is there a practice of benchmarking projects in relation to projects developed by similar companies? | 0 |
| **03** | Is there practice of contracting consultancies to identify and evaluate project maturity for successful projects (IPA - Independent Project Analysis, etc.)? | 0 |
| **04** | Is there practice of hiring consultancies in the evaluation and optimization of budget? | 7 |
| **05** | Is there a formal process of closing contracts open during the project? | 0 |
| **06** | Is there a formal process of collecting the lessons learned during and at the end of the project? | 7 |
| **07** | Is there a formal process of cost control through tools like EVA - Value Added Estimation? | 0 |
| **08** | Is there a formal operational risk analysis and survey process (HAZOP - Hazard and Operability Studies)? | 0 |

## *Appendix*

# *Table 58. Engineering and geology*

| Id. | Question | Score |
|---|---|---|
| 01 | Is there a formalized process that every project developed in the organization should go through the phases: conceptual, basic and detailed? | 7 |
| 02 | Is every project being run by the organization going through or going through the conceptual, basic, and detailed engineering phases? | 7 |
| 03 | Is there a formalized process for the drilling and auditing of mineral resources for engineering? | 0 |
| 04 | Is there a formal process that all conceptual engineering should begin with the first results of exploratory surveys? | 0 |
| 05 | Is there a formalization and submission of conceptual engineering information to the Environmental area for environmental licensing? | |
| 06 | Is all Conceptual Engineering developed based on a mapping of biotopes (areas impacted by the environment) and surface? | 7 |
| 07 | Are all process tests (Technological Development) formally sent to Engineering to start Basic Engineering? | 0 |
| 08 | Is there a formalized process for reviewing engineering quality (Peer Review) before moving from one phase to another (e.g., Conceptual Engineering for Basic Engineering)? | 0 |
| 09 | Is there a formalized engineering development culture in 3D technology? | 0 |
| 10 | Is there a process of evaluation, prior to the beginning of the Detailed Engineering, on the detailing of the project through the modularization of the construction? (Construction in Modules) | 7 |
| 11 | Is there the process of generating a Master Plan for the project after Conceptual Engineering? | 0 |
| 12 | If yes for answer 11, is there a formalized process of freezing this master plan and communicating it to all project stakeholders? | 7 |
| 13 | Is there a formalized process of generation of Descriptive Memorials, Spreadsheets, Data Sheet and Technical Specifications for the Supplies area? | 0 |
| 14 | Is there a formalized process of participation of the Supplies area during the evolution of the Basic Engineering for Detailed Engineering to determine the type of contracting of the Supply, Assembly and Implementation of the project? | 0 |
| 15 | Is there a formal process of realizing and studying tradeoffs for the lowest cost for utilities (water, gas, electricity, logistics, etc.)? | 7 |
| 16 | Is there a formal process of realization and study of opportunities (tradeoffs) regarding the location of the project (surface workers, communities, logistics, etc.)? | 7 |
| 17 | Is there a formal process for evaluating and validating each engineering phase (Conceptual, Basic, and Detailed) before moving from one to the other (Front End Loading Portions of Independent Project Analysis)? | 0 |
| 18 | At the end of the Project's Conceptual Engineering, is there a formal process to generate a Prefeasibility Study? | 0 |
| 19 | At the end of the Basic Engineering of the Project, is there a formal process to generate a Feasibility Study of the project? | 0 |
| 20 | During the execution of the Conceptual Engineering, is there a formal process of identification and first contacts with impacted landowners as well as local communities? | 7 |
| 21 | During the execution of Conceptual Engineering, is there a formal topographic survey of the region as well as the generation of orthophotos for the project's master plan? | 7 |
| 22 | After Conceptual Engineering, is there a formal process of generating investment estimates at a precision level of -15% to + 25%? | 7 |
| 23 | After the Basic Engineering, is there a formal process of generating the investment estimates at a precision level of -10% to + 10%? | 7 |
| 24 | Is there a formal closed-end planning process for the project after its useful life? | 0 |
| 25 | Is there a formal process for identifying VIPs - Value Improving Practices? | 0 |

*Table 59. Organizational structure*

| Id. | Question | Score |
|---|---|---|
| **01** | When a project starts, is there a project manager assigned by the organization to lead the project? | 7 |
| **02** | Is there an area focused on project planning? | 7 |
| **03** | Is there a formalized position of project manager within the organization? | 0 |
| **04** | Is there formalization of the Project Manager's power to similar functional positions (HR Manager, Operations Manager, etc.)? | 0 |
| **05** | Is there formalization of areas within Project management such as: Planning, PMO, Cost Control, etc.? | 0 |
| **06** | Are all internal areas of the company formally communicated from the beginning of the project and are their representatives formally assigned to the project? | 0 |

*Table 60. People management*

| Id. | Question | Score |
|---|---|---|
| **01** | Is there a formal process of developing the project manager's career and team? | 7 |
| **02** | Do follow-up meetings have a schedule and a fixed format? | 7 |
| **03** | Does each project member receive their work with instructions and in advance? | 7 |
| **04** | Is there a formal process of variable compensation for project team performance? | 7 |
| **05** | Is there a formal process for evaluating project team performance? | 7 |
| **06** | Is there a formal organization chart of the project team, delegations, etc.? | 7 |
| **07** | Are all the human resources required for the project formally sent to the HR area to arrange recruitment? | 7 |
| **08** | Are all interpersonal skills of the project team managed during the project? | 7 |
| **09** | Are all conflicts between the project team formally addressed? | 0 |

*Table 61. Project management office*

| Id. | Question | Score |
|---|---|---|
| 01 | Is there a flow for the arrival in your area of the information about the existence of a project or a contract? | 7 |
| 02 | Is there a dedication of the professionals to the management of the projects? | 7 |
| 03 | Is there any tool for estimating effort (time) and project costs? | 0 |
| 04 | Is there any tool for estimating project resources? | 7 |
| 05 | Is there any tool to formalize the client's acceptance? | 0 |
| 06 | Is there an established process for formal acceptance of each deliverable of the project? | 7 |
| 07 | For the development of project documentation, is there a repository of past project templates for reuse in new projects? | 7 |
| 08 | Is there an area focused on the physical and financial control of the project? | 7 |
| 09 | Is there a visibility room, a mural or even an intranet site where everyone can access projects in a simple way? | 0 |
| 10 | In the case of geographically dispersed teams, are the procedures / technologies used to integrate them adequate? | 0 |
| 11 | Are there managerial reports with a previously defined format agreed upon with the client? | 0 |
| 12 | Is there a formalized process for the closing of a project / contract? | 0 |
| 13 | Is there a formal process for the provision of information in the closing phase (finalization of the work)? | 0 |
| 14 | There is an area focused on the control and support of the project (PMO or Project Office) | 7 |

*Table 62. Stakeholder management*

| Id. | Question | Score |
|---|---|---|
| 01 | Is there a formalized process of mapping and managing the company's internal stakeholders? | 7 |
| 02 | Is there a formalized process for managing stakeholder expectations of the project during its execution? | 7 |
| 03 | Are formal interviews conducted to identify project stakeholders? | 0 |
| 04 | Are all potential impacts caused by each stakeholder formally assessed and managed during and after the project? | 7 |
| 05 | Is there a formal process of ranking and prioritizing the degree of influence of each stakeholder on the project? | 0 |
| 06 | Is there a formalized process of classification of stakeholder groups? | 0 |
| 07 | Is there a formal process of simulating the reactions of each stakeholder to the various situations that may occur during and after the project? | 7 |
| 08 | Is there a formalized process of gathering and filing the information of each stakeholder such as: name, address, telephone, etc.? | 7 |
| 09 | Is there a formalized Communication Plan capable of identifying what kind of information, when, how and for which stakeholder to send? | 7 |

*Table 63. Environment management*

| Id. | Question | Score |
|---|---|---|
| 01 | Is there a formalized process of beginning environmental studies with the beginning of Conceptual Engineering? | 7 |
| 02 | Is there a formalized process for mapping biotopes (areas of environmental impact) to aid in the development of Conceptual Engineering? | 7 |
| 03 | Is there a formalization process of mapping the surface properties impacted by the project during the Environmental Licensing and Conceptual Engineering? | 7 |
| 04 | Is there a formal process of generating documentation for Environmental Impact Studies and Environmental Impact Report by the Engineering area? | 7 |
| 05 | Is there a formal process that synchronizes the development of Engineering and Environment activities? (Generation of Documentation for Environmental Licensing)? | 7 |
| 06 | Is there a formalized area for institutional relations with government agencies? | 7 |
| 07 | Is there a formalized area for relationship with the communities impacted by the project? | 7 |
| 08 | Is there a formalized integration of the internal areas of the company: Engineering, Environment, Institutional Relationship and Social Responsibility? | 0 |
| 09 | Is there a formal process to monitor the release of Pre-Installation, Installation and Operation licenses from government agencies? | 0 |

## *Appendix*

# *Table 64. Project management*

| Id. | Question | Score |
|---|---|---|
| **01** | Is there any document to formalize the existence of the project in your area? | 7 |
| **02** | Does the project existence document adequately describe the project output? | 0 |
| **03** | Is there a document that can document the assumptions (prerequisites necessary for you to complete the work) and constraints (dates imposed by the client, limitation of funding, limitation of resources) of the project? | 7 |
| **04** | Is there a Project Analytical Framework (EAP) or WBS? | 0 |
| **05** | Is there any tool to formalize the client's acceptance? | 7 |
| **06** | Is there a baseline against which project progress, costs and deadlines are measured? | 7 |
| **07** | Is there an established process for project change control? | 7 |
| **08** | Is there a process of identifying predecessors and successors of project activities? | 0 |
| **09** | Is there a separation of project activities according to each schedule of the resource to be used (business days, holidays, etc.)? | 0 |
| **10** | Is there a process for identifying and prioritizing the critical path of the project? | 7 |
| **11** | Is there a baseline recording of the project at the end of the schedule development? | 7 |
| **12** | In developing the schedule, do all activities have resources? | 0 |
| **13** | Is there a formalization process of generating provisions (contingencies) for project costs? | 0 |
| **14** | Is the process of formalizing a matrix of responsibilities for the project? | 0 |
| **15** | Is there a formalized process of generating and transmitting project information during its execution (Communications Plan)? | 0 |
| **16** | Is there a mapping that details who needs the information, when it will be needed, how it should be defined, and by whom? | 0 |
| **17** | Is there a place with phones, addresses, e-mails and contacts from the entire team and involved? | 0 |
| **18** | Is there a formalized process for defining the schedule of project meetings? | 0 |
| **20** | Are the meetings pre-scheduled from the start and on the agenda of the entire team? | 0 |
| **21** | Does each team member know what information to bring to the follow-up meeting? | 0 |
| **22** | Is there a formal project budgeting process by grouping estimates of activities, resources, hiring, contingencies, etc.? | 0 |
| **23** | Is there a formalized risk identification process for the project? | 7 |
| **24** | Is there a qualitative and quantitative project risk assessment process? | 7 |
| **25** | Is there a project risk prioritization process? | 7 |
| **26** | Is there a process for planning the project's risk responses? | 7 |
| **27** | Is there a process to identify who is responsible for the project risk? | 7 |
| **28** | Is there a risk management and monitoring tool for the project? | 0 |
| **29** | Is there a process for identifying project product requirements? | 0 |
| **30** | Are the project product requirements translated into the work required to develop this product? | 0 |
| **31** | Is there a document that can formalize the work of the project (Scope Statement)? | 7 |
| **32** | Is there a formalized audit plan for quality requirements for the project (Quality Plan)? | 0 |
| **33** | Is there formalized planning for the provision and contracting of project resources (engineering, equipment, etc.)? | 0 |
| **34** | Is there a formalized contracting cycle for large project deliverables (Bid Cycle or Hiring Cycle)? | 7 |

*Table 65. Performance and results*

| Id. | Question | Score |
|---|---|---|
| **01** | Are deadlines met as planned? | 0 |
| **02** | Are the quality of the delivered projects according to the customer's requirements? | 0 |
| **03** | Are project costs met as planned? | 0 |
| **04** | Is there a monitoring of customer satisfaction with the project product? | 0 |
| **05** | Is there a process for evaluating the performance of the project manager in the conduct of the project? | 0 |
| **06** | Is there a project team evaluation process? | 7 |

*Table 66. Methods and practices*

| Id. | Question | Score |
|---|---|---|
| **01** | Before the start of the project is a feasibility study for the investment carried out? | 7 |
| **02** | Is there a practice of benchmarking projects in relation to projects developed by similar companies? | 0 |
| **03** | Is there practice of contracting consultancies to identify and evaluate project maturity for successful projects (IPA - Independent Project Analysis, etc.)? | 0 |
| **04** | Is there practice of hiring consultancies in the evaluation and optimization of budget? | 0 |
| **05** | Is there a formal process of closing contracts open during the project? | 7 |
| **06** | Is there a formal process of collecting the lessons learned during and at the end of the project? | 7 |
| **07** | Is there a formal process of cost control through tools like EVA - Value Added Estimation? | 0 |
| **08** | Is there a formal operational risk analysis and survey process (HAZOP - Hazard and Operability Studies)? | 0 |

## Table 67. Engineering and geology

| Id. | Question | Score |
|---|---|---|
| 01 | Is there a formalized process that every project developed in the organization should go through the phases: conceptual, basic and detailed? | 0 |
| 02 | Is every project being run by the organization going through or going through the conceptual, basic, and detailed engineering phases? | 0 |
| 03 | Is there a formalized process for the drilling and auditing of mineral resources for engineering? | 0 |
| 04 | Is there a formal process that all conceptual engineering should begin with the first results of exploratory surveys? | 0 |
| 05 | Is there a formalization and submission of conceptual engineering information to the Environmental area for environmental licensing? | 0 |
| 06 | Is all Conceptual Engineering developed based on a mapping of biotopes (areas impacted by the environment) and surface? | 0 |
| 07 | Are all process tests (Technological Development) formally sent to Engineering to start Basic Engineering? | 0 |
| 08 | Is there a formalized process for reviewing engineering quality (Peer Review) before moving from one phase to another (e.g., Conceptual Engineering for Basic Engineering)? | 0 |
| 09 | Is there a formalized engineering development culture in 3D technology? | 0 |
| 10 | Is there a process of evaluation, prior to the beginning of the Detailed Engineering, on the detailing of the project through the modularization of the construction? (Construction in Modules) | 0 |
| 11 | Is there the process of generating a Master Plan for the project after Conceptual Engineering? | 7 |
| 12 | If yes for answer 11, is there a formalized process of freezing this master plan and communicating it to all project stakeholders? | 0 |
| 13 | Is there a formalized process of generation of Descriptive Memorials, Spreadsheets, Data Sheet and Technical Specifications for the Supplies area? | 7 |
| 14 | Is there a formalized process of participation of the Supplies area during the evolution of the Basic Engineering for Detailed Engineering to determine the type of contracting of the Supply, Assembly and Implementation of the project? | 0 |
| 15 | Is there a formal process of realizing and studying tradeoffs for the lowest cost for utilities (water, gas, electricity, logistics, etc.)? | 0 |
| 16 | Is there a formal process of realization and study of opportunities (tradeoffs) regarding the location of the project (surface workers, communities, logistics, etc.)? | 7 |
| 17 | Is there a formal process for evaluating and validating each engineering phase (Conceptual, Basic, and Detailed) before moving from one to the other (Front End Loading Portions of Independent Project Analysis)? | 0 |
| 18 | At the end of the Project's Conceptual Engineering, is there a formal process to generate a Prefeasibility Study? | 7 |
| 19 | At the end of the Basic Engineering of the Project, is there a formal process to generate a Feasibility Study of the project? | 0 |
| 20 | During the execution of the Conceptual Engineering, is there a formal process of identification and first contacts with impacted landowners as well as local communities? | 0 |
| 21 | During the execution of Conceptual Engineering, is there a formal topographic survey of the region as well as the generation of orthophotos for the project's master plan? | 7 |
| 22 | After Conceptual Engineering, is there a formal process of generating investment estimates at a precision level of -15% to + 25%? | 0 |
| 23 | After the Basic Engineering, is there a formal process of generating the investment estimates at a precision level of -10% to + 10%? | 7 |
| 24 | Is there a formal closed-end planning process for the project after its useful life? | 0 |
| 25 | Is there a formal process for identifying VIPs - Value Improving Practices? | 0 |

*Table 68. Organizational structure*

| Id. | Question | Score |
|---|---|---|
| **01** | When a project starts, is there a project manager assigned by the organization to lead the project? | 0 |
| **02** | Is there an area focused on project planning? | 0 |
| **03** | Is there a formalized position of project manager within the organization? | 0 |
| **04** | Is there formalization of the Project Manager's power to similar functional positions (HR Manager, Operations Manager, etc.)? | 7 |
| **05** | Is there formalization of areas within Project management such as: Planning, PMO, Cost Control, etc.? | 7 |
| **06** | Are all internal areas of the company formally communicated from the beginning of the project and are their representatives formally assigned to the project? | 0 |

*Table 69. People management*

| Id. | Question | Score |
|---|---|---|
| **01** | Is there a formal process of developing the project manager's career and team? | 0 |
| **02** | Do follow-up meetings have a schedule and a fixed format? | 0 |
| **03** | Does each project member receive their work with instructions and in advance? | 0 |
| **04** | Is there a formal process of variable compensation for project team performance? | 0 |
| **05** | Is there a formal process for evaluating project team performance? | 7 |
| **06** | Is there a formal organization chart of the project team, delegations, etc.? | 0 |
| **07** | Are all the human resources required for the project formally sent to the HR area to arrange recruitment? | 0 |
| **08** | Are all interpersonal skills of the project team managed during the project? | 0 |
| **09** | Are all conflicts between the project team formally addressed? | 0 |

## Table 70. Project management office

| Id. | Question | Score |
|---|---|---|
| 01 | Is there a flow for the arrival in your area of the information about the existence of a project or a contract? | 0 |
| 02 | Is there a dedication of the professionals to the management of the projects? | 7 |
| 03 | Is there any tool for estimating effort (time) and project costs? | 0 |
| 04 | Is there any tool for estimating project resources? | 0 |
| 05 | Is there any tool to formalize the client's acceptance? | 7 |
| 06 | Is there an established process for formal acceptance of each deliverable of the project? | 0 |
| 07 | For the development of project documentation, is there a repository of past project templates for reuse in new projects? | 0 |
| 08 | Is there an area focused on the physical and financial control of the project? | 0 |
| 09 | Is there a visibility room, a mural or even an intranet site where everyone can access projects in a simple way? | 0 |
| 10 | In the case of geographically dispersed teams, are the procedures / technologies used to integrate them adequate? | 0 |
| 11 | Are there managerial reports with a previously defined format agreed upon with the client? | 7 |
| 12 | Is there a formalized process for the closing of a project / contract? | 0 |
| 13 | Is there a formal process for the provision of information in the closing phase (finalization of the work)? | 0 |
| 14 | There is an area focused on the control and support of the project (PMO or Project Office) | 7 |

## Table 71. Stakeholder management

| Id. | Question | Score |
|---|---|---|
| 01 | Is there a formalized process of mapping and managing the company's internal stakeholders? | 0 |
| 02 | Is there a formalized process for managing stakeholder expectations of the project during its execution? | 7 |
| 03 | Are formal interviews conducted to identify project stakeholders? | 0 |
| 04 | Are all potential impacts caused by each stakeholder formally assessed and managed during and after the project? | 7 |
| 05 | Is there a formal process of ranking and prioritizing the degree of influence of each stakeholder on the project? | 0 |
| 06 | Is there a formalized process of classification of stakeholder groups? | 7 |
| 07 | Is there a formal process of simulating the reactions of each stakeholder to the various situations that may occur during and after the project? | 0 |
| 08 | Is there a formalized process of gathering and filing the information of each stakeholder such as: name, address, telephone, etc.? | 7 |
| 09 | Is there a formalized Communication Plan capable of identifying what kind of information, when, how and for which stakeholder to send? | 0 |

*Table 72. Environment management*

| Id. | Question | Score |
|---|---|---|
| **01** | Is there a formalized process of beginning environmental studies with the beginning of Conceptual Engineering? | 0 |
| **02** | Is there a formalized process for mapping biotopes (areas of environmental impact) to aid in the development of Conceptual Engineering? | 0 |
| **03** | Is there a formalization process of mapping the surface properties impacted by the project during the Environmental Licensing and Conceptual Engineering? | 0 |
| **04** | Is there a formal process of generating documentation for Environmental Impact Studies and Environmental Impact Report by the Engineering area? | 7 |
| **05** | Is there a formal process that synchronizes the development of Engineering and Environment activities? (Generation of Documentation for Environmental Licensing)? | 7 |
| **06** | Is there a formalized area for institutional relations with government agencies? | 7 |
| **07** | Is there a formalized area for relationship with the communities impacted by the project? | 7 |
| **08** | Is there a formalized integration of the internal areas of the company: Engineering, Environment, Institutional Relationship and Social Responsibility? | 0 |
| **09** | Is there a formal process to monitor the release of Pre-Installation, Installation and Operation licenses from government agencies? | 7 |

## Table 73. Project management

| Id. | Question | Score |
|---|---|---|
| 01 | Is there any document to formalize the existence of the project in your area? | 0 |
| 02 | Does the project existence document adequately describe the project output? | 0 |
| 03 | Is there a document that can document the assumptions (prerequisites necessary for you to complete the work) and constraints (dates imposed by the client, limitation of funding, limitation of resources) of the project? | 7 |
| 04 | Is there a Project Analytical Framework (EAP) or WBS? | 0 |
| 05 | Is there any tool to formalize the client's acceptance? | 0 |
| 06 | Is there a baseline against which project progress, costs and deadlines are measured? | 0 |
| 07 | Is there an established process for project change control? | 7 |
| 08 | Is there a process of identifying predecessors and successors of project activities? | 7 |
| 09 | Is there a separation of project activities according to each schedule of the resource to be used (business days, holidays, etc.)? | 7 |
| 10 | Is there a process for identifying and prioritizing the critical path of the project? | 0 |
| 11 | Is there a baseline recording of the project at the end of the schedule development? | 0 |
| 12 | In developing the schedule, do all activities have resources? | 0 |
| 13 | Is there a formalization process of generating provisions (contingencies) for project costs? | 0 |
| 14 | Is the process of formalizing a matrix of responsibilities for the project? | 0 |
| 15 | Is there a formalized process of generating and transmitting project information during its execution (Communications Plan)? | 0 |
| 16 | Is there a mapping that details who needs the information, when it will be needed, how it should be defined, and by whom? | 0 |
| 17 | Is there a place with phones, addresses, e-mails and contacts from the entire team and involved? | 7 |
| 18 | Is there a formalized process for defining the schedule of project meetings? | 7 |
| 20 | Are the meetings pre-scheduled from the start and on the agenda of the entire team? | 7 |
| 21 | Does each team member know what information to bring to the follow-up meeting? | 0 |
| 22 | Is there a formal project budgeting process by grouping estimates of activities, resources, hiring, contingencies, etc.? | 7 |
| 23 | Is there a formalized risk identification process for the project? | 0 |
| 24 | Is there a qualitative and quantitative project risk assessment process? | 0 |
| 25 | Is there a project risk prioritization process? | 0 |
| 26 | Is there a process for planning the project's risk responses? | 0 |
| 27 | Is there a process to identify who is responsible for the project risk? | 0 |
| 28 | Is there a risk management and monitoring tool for the project? | 0 |
| 29 | Is there a process for identifying project product requirements? | 0 |
| 30 | Are the project product requirements translated into the work required to develop this product? | 0 |
| 31 | Is there a document that can formalize the work of the project (Scope Statement)? | 7 |
| 32 | Is there a formalized audit plan for quality requirements for the project (Quality Plan)? | 0 |
| 33 | Is there formalized planning for the provision and contracting of project resources (engineering, equipment, etc.)? | 0 |
| 34 | Is there a formalized contracting cycle for large project deliverables (Bid Cycle or Hiring Cycle)? | 7 |

*Table 74. Performance and results*

| Id. | Quesion | Score |
|---|---|---|
| **01** | Are deadlines met as planned? | 0 |
| **02** | Are the quality of the delivered projects according to the customer's requirements? | 7 |
| **03** | Are project costs met as planned? | 7 |
| **04** | Is there a monitoring of customer satisfaction with the project product? | 0 |
| **05** | Is there a process for evaluating the performance of the project manager in the conduct of the project? | 0 |
| **06** | Is there a project team evaluation process? | 0 |

*Table 75. Methods and practices*

| Id. | Question | Score |
|---|---|---|
| **01** | Before the start of the project is a feasibility study for the investment carried out? | 7 |
| **02** | Is there a practice of benchmarking projects in relation to projects developed by similar companies? | 0 |
| **03** | Is there practice of contracting consultancies to identify and evaluate project maturity for successful projects (IPA - Independent Project Analysis, etc.)? | 7 |
| **04** | Is there practice of hiring consultancies in the evaluation and optimization of budget? | 7 |
| **05** | Is there a formal process of closing contracts open during the project? | 0 |
| **06** | Is there a formal process of collecting the lessons learned during and at the end of the project? | 0 |
| **07** | Is there a formal process of cost control through tools like EVA - Value Added Estimation? | 0 |
| **08** | Is there a formal operational risk analysis and survey process (HAZOP - Hazard and Operability Studies)? | 0 |

# Table 76. Engineering and geology

| Id. | Question | Score |
|---|---|---|
| 01 | Is there a formalized process that every project developed in the organization should go through the phases: conceptual, basic and detailed? | 7 |
| 02 | Is every project being run by the organization going through or going through the conceptual, basic, and detailed engineering phases? | 0 |
| 03 | Is there a formalized process for the drilling and auditing of mineral resources for engineering? | 0 |
| 04 | Is there a formal process that all conceptual engineering should begin with the first results of exploratory surveys? | 0 |
| 05 | Is there a formalization and submission of conceptual engineering information to the Environmental area for environmental licensing? | 0 |
| 06 | Is all Conceptual Engineering developed based on a mapping of biotopes (areas impacted by the environment) and surface? | 0 |
| 07 | Are all process tests (Technological Development) formally sent to Engineering to start Basic Engineering? | 0 |
| 08 | Is there a formalized process for reviewing engineering quality (Peer Review) before moving from one phase to another (e.g., Conceptual Engineering for Basic Engineering)? | 0 |
| 09 | Is there a formalized engineering development culture in 3D technology? | 0 |
| 10 | Is there a process of evaluation, prior to the beginning of the Detailed Engineering, on the detailing of the project through the modularization of the construction? (Construction in Modules) | 0 |
| 11 | Is there the process of generating a Master Plan for the project after Conceptual Engineering? | 0 |
| 12 | If yes for answer 11, is there a formalized process of freezing this master plan and communicating it to all project stakeholders? | |
| 13 | Is there a formalized process of generation of Descriptive Memorials, Spreadsheets, Data Sheet and Technical Specifications for the Supplies area? | 7 |
| 14 | Is there a formalized process of participation of the Supplies area during the evolution of the Basic Engineering for Detailed Engineering to determine the type of contracting of the Supply, Assembly and Implementation of the project? | 7 |
| 15 | Is there a formal process of realizing and studying tradeoffs for the lowest cost for utilities (water, gas, electricity, logistics, etc.)? | 7 |
| 16 | Is there a formal process of realization and study of opportunities (tradeoffs) regarding the location of the project (surface workers, communities, logistics, etc.)? | 7 |
| 17 | Is there a formal process for evaluating and validating each engineering phase (Conceptual, Basic, and Detailed) before moving from one to the other (Front End Loading Portions of Independent Project Analysis)? | 0 |
| 18 | At the end of the Project's Conceptual Engineering, is there a formal process to generate a Prefeasibility Study? | 0 |
| 19 | At the end of the Basic Engineering of the Project, is there a formal process to generate a Feasibility Study of the project? | 0 |
| 20 | During the execution of the Conceptual Engineering, is there a formal process of identification and first contacts with impacted landowners as well as local communities? | 7 |
| 21 | During the execution of Conceptual Engineering, is there a formal topographic survey of the region as well as the generation of orthophotos for the project's master plan? | 7 |
| 22 | After Conceptual Engineering, is there a formal process of generating investment estimates at a precision level of -15% to + 25%? | 7 |
| 23 | After the Basic Engineering, is there a formal process of generating the investment estimates at a precision level of -10% to + 10%? | 0 |
| 24 | Is there a formal closed-end planning process for the project after its useful life? | 0 |
| 25 | Is there a formal process for identifying VIPs - Value Improving Practices? | 0 |

*Table 77. Organizational structure*

| Id. | Question | Score |
|---|---|---|
| **01** | When a project starts, is there a project manager assigned by the organization to lead the project? | 7 |
| **02** | Is there an area focused on project planning? | 7 |
| **03** | Is there a formalized position of project manager within the organization? | 7 |
| **04** | Is there formalization of the Project Manager's power to similar functional positions (HR Manager, Operations Manager, etc.)? | 0 |
| **05** | Is there formalization of areas within Project management such as: Planning, PMO, Cost Control, etc.? | 7 |
| **06** | Are all internal areas of the company formally communicated from the beginning of the project and are their representatives formally assigned to the project? | 0 |

*Table 78. People management*

| Id. | Question | Score |
|---|---|---|
| **01** | Is there a formal process of developing the project manager's career and team? | 0 |
| **02** | Do follow-up meetings have a schedule and a fixed format? | 7 |
| **03** | Does each project member receive their work with instructions and in advance? | 0 |
| **04** | Is there a formal process of variable compensation for project team performance? | 7 |
| **05** | Is there a formal process for evaluating project team performance? | 7 |
| **06** | Is there a formal organization chart of the project team, delegations, etc.? | 0 |
| **07** | Are all the human resources required for the project formally sent to the HR area to arrange recruitment? | 0 |
| **08** | Are all interpersonal skills of the project team managed during the project? | 0 |
| **09** | Are all conflicts between the project team formally addressed? | 0 |

Table 79. Project management office

| Id. | Question | Score |
|---|---|---|
| 01 | Is there a flow for the arrival in your area of the information about the existence of a project or a contract? | 0 |
| 02 | Is there a dedication of the professionals to the management of the projects? | 7 |
| 03 | Is there any tool for estimating effort (time) and project costs? | 7 |
| 04 | Is there any tool for estimating project resources? | 0 |
| 05 | Is there any tool to formalize the client's acceptance? | 7 |
| 06 | Is there an established process for formal acceptance of each deliverable of the project? | 0 |
| 07 | For the development of project documentation, is there a repository of past project templates for reuse in new projects? | 0 |
| 08 | Is there an area focused on the physical and financial control of the project? | 0 |
| 09 | Is there a visibility room, a mural or even an intranet site where everyone can access projects in a simple way? | 0 |
| 10 | In the case of geographically dispersed teams, are the procedures / technologies used to integrate them adequate? | 7 |
| 11 | Are there managerial reports with a previously defined format agreed upon with the client? | 7 |
| 12 | Is there a formalized process for the closing of a project / contract? | 0 |
| 13 | Is there a formal process for the provision of information in the closing phase (finalization of the work)? | 0 |
| 14 | There is an area focused on the control and support of the project (PMO or Project Office) | 7 |

Table 80. Stakeholder management

| Id. | Question | Score |
|---|---|---|
| 01 | Is there a formalized process of mapping and managing the company's internal stakeholders? | 7 |
| 02 | Is there a formalized process for managing stakeholder expectations of the project during its execution? | 0 |
| 03 | Are formal interviews conducted to identify project stakeholders? | 0 |
| 04 | Are all potential impacts caused by each stakeholder formally assessed and managed during and after the project? | 7 |
| 05 | Is there a formal process of ranking and prioritizing the degree of influence of each stakeholder on the project? | 0 |
| 06 | Is there a formalized process of classification of stakeholder groups? | 0 |
| 07 | Is there a formal process of simulating the reactions of each stakeholder to the various situations that may occur during and after the project? | 7 |
| 08 | Is there a formalized process of gathering and filing the information of each stakeholder such as: name, address, telephone, etc.? | 0 |
| 09 | Is there a formalized Communication Plan capable of identifying what kind of information, when, how and for which stakeholder to send? | 0 |

*Table 81. Environment management*

| Id. | Question | Score |
|---|---|---|
| 01 | Is there a formalized process of beginning environmental studies with the beginning of Conceptual Engineering? | 7 |
| 02 | Is there a formalized process for mapping biotopes (areas of environmental impact) to aid in the development of Conceptual Engineering? | 7 |
| 03 | Is there a formalization process of mapping the surface properties impacted by the project during the Environmental Licensing and Conceptual Engineering? | 7 |
| 04 | Is there a formal process of generating documentation for Environmental Impact Studies and Environmental Impact Report by the Engineering area? | 7 |
| 05 | Is there a formal process that synchronizes the development of Engineering and Environment activities? (Generation of Documentation for Environmental Licensing)? | 0 |
| 06 | Is there a formalized area for institutional relations with government agencies? | 7 |
| 07 | Is there a formalized area for relationship with the communities impacted by the project? | 7 |
| 08 | Is there a formalized integration of the internal areas of the company: Engineering, Environment, Institutional Relationship and Social Responsibility? | 0 |
| 09 | Is there a formal process to monitor the release of Pre-Installation, Installation and Operation licenses from government agencies? | 7 |

# Table 82. Project management

| Id. | Question | Score |
|---|---|---|
| 01 | Is there any document to formalize the existence of the project in your area? | 7 |
| 02 | Does the project existence document adequately describe the project output? | 7 |
| 03 | Is there a document that can document the assumptions (prerequisites necessary for you to complete the work) and constraints (dates imposed by the client, limitation of funding, limitation of resources) of the project? | 7 |
| 04 | Is there a Project Analytical Framework (EAP) or WBS? | 7 |
| 05 | Is there any tool to formalize the client's acceptance? | 7 |
| 06 | Is there a baseline against which project progress, costs and deadlines are measured? | 7 |
| 07 | Is there an established process for project change control? | 7 |
| 08 | Is there a process of identifying predecessors and successors of project activities? | 7 |
| 09 | Is there a separation of project activities according to each schedule of the resource to be used (business days, holidays, etc.)? | 0 |
| 10 | Is there a process for identifying and prioritizing the critical path of the project? | 7 |
| 11 | Is there a baseline recording of the project at the end of the schedule development? | 7 |
| 12 | In developing the schedule, do all activities have resources? | 0 |
| 13 | Is there a formalization process of generating provisions (contingencies) for project costs? | 0 |
| 14 | Is the process of formalizing a matrix of responsibilities for the project? | 7 |
| 15 | Is there a formalized process of generating and transmitting project information during its execution (Communications Plan)? | 7 |
| 16 | Is there a mapping that details who needs the information, when it will be needed, how it should be defined, and by whom? | 0 |
| 17 | Is there a place with phones, addresses, e-mails and contacts from the entire team and involved? | 7 |
| 18 | Is there a formalized process for defining the schedule of project meetings? | 0 |
| 20 | Are the meetings pre-scheduled from the start and on the agenda of the entire team? | 7 |
| 21 | Does each team member know what information to bring to the follow-up meeting? | 0 |
| 22 | Is there a formal project budgeting process by grouping estimates of activities, resources, hiring, contingencies, etc.? | 7 |
| 23 | Is there a formalized risk identification process for the project? | 0 |
| 24 | Is there a qualitative and quantitative project risk assessment process? | 0 |
| 25 | Is there a project risk prioritization process? | 0 |
| 26 | Is there a process for planning the project's risk responses? | 0 |
| 27 | Is there a process to identify who is responsible for the project risk? | 0 |
| 28 | Is there a risk management and monitoring tool for the project? | 0 |
| 29 | Is there a process for identifying project product requirements? | 7 |
| 30 | Are the project product requirements translated into the work required to develop this product? | 7 |
| 31 | Is there a document that can formalize the work of the project (Scope Statement)? | 7 |
| 32 | Is there a formalized audit plan for quality requirements for the project (Quality Plan)? | 0 |
| 33 | Is there formalized planning for the provision and contracting of project resources (engineering, equipment, etc.)? | 7 |
| 34 | Is there a formalized contracting cycle for large project deliverables (Bid Cycle or Hiring Cycle)? | 7 |

*Table 83. Performance and results*

| Id. | Question | Score |
|---|---|---|
| **01** | Are deadlines met as planned? | 0 |
| **02** | Are the quality of the delivered projects according to the customer's requirements? | 7 |
| **03** | Are project costs met as planned? | 0 |
| **04** | Is there a monitoring of customer satisfaction with the project product? | 7 |
| **05** | Is there a process for evaluating the performance of the project manager in the conduct of the project? | 0 |
| **06** | Is there a project team evaluation process? | 0 |

*Table 84. Methods and practices*

| Id. | Question | Score |
|---|---|---|
| **01** | Before the start of the project is a feasibility study for the investment carried out? | 7 |
| **02** | Is there a practice of benchmarking projects in relation to projects developed by similar companies? | 7 |
| **03** | Is there practice of contracting consultancies to identify and evaluate project maturity for successful projects (IPA - Independent Project Analysis, etc.)? | 0 |
| **04** | Is there practice of hiring consultancies in the evaluation and optimization of budget? | 0 |
| **05** | Is there a formal process of closing contracts open during the project? | 7 |
| **06** | Is there a formal process of collecting the lessons learned during and at the end of the project? | 0 |
| **07** | Is there a formal process of cost control through tools like EVA - Value Added Estimation? | 7 |
| **08** | Is there a formal operational risk analysis and survey process (HAZOP - Hazard and Operability Studies)? | 0 |

## Appendix

*Table 85. Engineering and geology*

| Id. | Question | Score |
|---|---|---|
| 01 | Is there a formalized process that every project developed in the organization should go through the phases: conceptual, basic and detailed? | 7 |
| 02 | Is every project being run by the organization going through or going through the conceptual, basic, and detailed engineering phases? | 7 |
| 03 | Is there a formalized process for the drilling and auditing of mineral resources for engineering? | 7 |
| 04 | Is there a formal process that all conceptual engineering should begin with the first results of exploratory surveys? | 7 |
| 05 | Is there a formalization and submission of conceptual engineering information to the Environmental area for environmental licensing? | |
| 06 | Is all Conceptual Engineering developed based on a mapping of biotopes (areas impacted by the environment) and surface? | 7 |
| 07 | Are all process tests (Technological Development) formally sent to Engineering to start Basic Engineering? | 7 |
| 08 | Is there a formalized process for reviewing engineering quality (Peer Review) before moving from one phase to another (e.g., Conceptual Engineering for Basic Engineering)? | 7 |
| 09 | Is there a formalized engineering development culture in 3D technology? | 0 |
| 10 | Is there a process of evaluation, prior to the beginning of the Detailed Engineering, on the detailing of the project through the modularization of the construction? (Construction in Modules) | 7 |
| 11 | Is there the process of generating a Master Plan for the project after Conceptual Engineering? | 7 |
| 12 | If yes for answer 11, is there a formalized process of freezing this master plan and communicating it to all project stakeholders? | 7 |
| 13 | Is there a formalized process of generation of Descriptive Memorials, Spreadsheets, Data Sheet and Technical Specifications for the Supplies area? | 7 |
| 14 | Is there a formalized process of participation of the Supplies area during the evolution of the Basic Engineering for Detailed Engineering to determine the type of contracting of the Supply, Assembly and Implementation of the project? | 7 |
| 15 | Is there a formal process of realizing and studying tradeoffs for the lowest cost for utilities (water, gas, electricity, logistics, etc.)? | |
| 16 | Is there a formal process of realization and study of opportunities (tradeoffs) regarding the location of the project (surface workers, communities, logistics, etc.)? | |
| 17 | Is there a formal process for evaluating and validating each engineering phase (Conceptual, Basic, and Detailed) before moving from one to the other (Front End Loading Portions of Independent Project Analysis)? | 0 |
| 18 | At the end of the Project's Conceptual Engineering, is there a formal process to generate a Prefeasibility Study? | 7 |
| 19 | At the end of the Basic Engineering of the Project, is there a formal process to generate a Feasibility Study of the project? | 7 |
| 20 | During the execution of the Conceptual Engineering, is there a formal process of identification and first contacts with impacted landowners as well as local communities? | |
| 21 | During the execution of Conceptual Engineering, is there a formal topographic survey of the region as well as the generation of orthophotos for the project's master plan? | 7 |
| 22 | After Conceptual Engineering, is there a formal process of generating investment estimates at a precision level of -15% to + 25%? | 0 |
| 23 | After the Basic Engineering, is there a formal process of generating the investment estimates at a precision level of -10% to + 10%? | 0 |
| 24 | Is there a formal closed-end planning process for the project after its useful life? | |
| 25 | Is there a formal process for identifying VIPs - Value Improving Practices? | 0 |

*Table 86. Organizational structure*

| Id. | Question | Score |
|---|---|---|
| 01 | When a project starts, is there a project manager assigned by the organization to lead the project? | 7 |
| 02 | Is there an area focused on project planning? | 0 |
| 03 | Is there a formalized position of project manager within the organization? | 7 |
| 04 | Is there formalization of the Project Manager's power to similar functional positions (HR Manager, Operations Manager, etc.)? | 0 |
| 05 | Is there formalization of areas within Project management such as: Planning, PMO, Cost Control, etc.? | 0 |
| 06 | Are all internal areas of the company formally communicated from the beginning of the project and are their representatives formally assigned to the project? | 7 |

*Table 87. People management*

| Id. | Question | Score |
|---|---|---|
| 01 | Is there a formal process of developing the project manager's career and team? | 0 |
| 02 | Do follow-up meetings have a schedule and a fixed format? | 7 |
| 03 | Does each project member receive their work with instructions and in advance? | 7 |
| 04 | Is there a formal process of variable compensation for project team performance? | 7 |
| 05 | Is there a formal process for evaluating project team performance? | 0 |
| 06 | Is there a formal organization chart of the project team, delegations, etc.? | 7 |
| 07 | Are all the human resources required for the project formally sent to the HR area to arrange recruitment? | 7 |
| 08 | Are all interpersonal skills of the project team managed during the project? | 0 |
| 09 | Are all conflicts between the project team formally addressed? | 0 |

*Table 88. Project management office*

| Id. | Question | Score |
|---|---|---|
| 01 | Is there a flow for the arrival in your area of the information about the existence of a project or a contract? | 7 |
| 02 | Is there a dedication of the professionals to the management of the projects? | 7 |
| 03 | Is there any tool for estimating effort (time) and project costs? | 7 |
| 04 | Is there any tool for estimating project resources? | 7 |
| 05 | Is there any tool to formalize the client's acceptance? | 7 |
| 06 | Is there an established process for formal acceptance of each deliverable of the project? | 7 |
| 07 | For the development of project documentation, is there a repository of past project templates for reuse in new projects? | 7 |
| 08 | Is there an area focused on the physical and financial control of the project? | 7 |
| 09 | Is there a visibility room, a mural or even an intranet site where everyone can access projects in a simple way? | 7 |
| 10 | In the case of geographically dispersed teams, are the procedures / technologies used to integrate them adequate? | 0 |
| 11 | Are there managerial reports with a previously defined format agreed upon with the client? | 7 |
| 12 | Is there a formalized process for the closing of a project / contract? | 7 |
| 13 | Is there a formal process for the provision of information in the closing phase (finalization of the work)? | 0 |
| 14 | There is an area focused on the control and support of the project (PMO or Project Office) | 0 |

*Table 89. Stakeholder management*

| Id. | Question | Score |
|---|---|---|
| 01 | Is there a formalized process of mapping and managing the company's internal stakeholders? | 7 |
| 02 | Is there a formalized process for managing stakeholder expectations of the project during its execution? | 7 |
| 03 | Are formal interviews conducted to identify project stakeholders? | 0 |
| 04 | Are all potential impacts caused by each stakeholder formally assessed and managed during and after the project? | 0 |
| 05 | Is there a formal process of ranking and prioritizing the degree of influence of each stakeholder on the project? | 0 |
| 06 | Is there a formalized process of classification of stakeholder groups? | 0 |
| 07 | Is there a formal process of simulating the reactions of each stakeholder to the various situations that may occur during and after the project? | 7 |
| 08 | Is there a formalized process of gathering and filing the information of each stakeholder such as: name, address, telephone, etc.? | 7 |
| 09 | Is there a formalized Communication Plan capable of identifying what kind of information, when, how and for which stakeholder to send? | 0 |

*Table 90. Environment management*

| Id. | Question | Score |
|---|---|---|
| 01 | Is there a formalized process of beginning environmental studies with the beginning of Conceptual Engineering? | |
| 02 | Is there a formalized process for mapping biotopes (areas of environmental impact) to aid in the development of Conceptual Engineering? | |
| 03 | Is there a formalization process of mapping the surface properties impacted by the project during the Environmental Licensing and Conceptual Engineering? | |
| 04 | Is there a formal process of generating documentation for Environmental Impact Studies and Environmental Impact Report by the Engineering area? | |
| 05 | Is there a formal process that synchronizes the development of Engineering and Environment activities? (Generation of Documentation for Environmental Licensing)? | |
| 06 | Is there a formalized area for institutional relations with government agencies? | |
| 07 | Is there a formalized area for relationship with the communities impacted by the project? | |
| 08 | Is there a formalized integration of the internal areas of the company: Engineering, Environment, Institutional Relationship and Social Responsibility? | |
| 09 | Is there a formal process to monitor the release of Pre-Installation, Installation and Operation licenses from government agencies? | |

## Appendix

*Table 91. Project management*

| Id. | Question | Score |
|---|---|---|
| 01 | Is there any document to formalize the existence of the project in your area? | 0 |
| 02 | Does the project existence document adequately describe the project output? | 0 |
| 03 | Is there a document that can document the assumptions (prerequisites necessary for you to complete the work) and constraints (dates imposed by the client, limitation of funding, limitation of resources) of the project? | 7 |
| 04 | Is there a Project Analytical Framework (EAP) or WBS? | 0 |
| 05 | Is there any tool to formalize the client's acceptance? | 7 |
| 06 | Is there a baseline against which project progress, costs and deadlines are measured? | 7 |
| 07 | Is there an established process for project change control? | 0 |
| 08 | Is there a process of identifying predecessors and successors of project activities? | 7 |
| 09 | Is there a separation of project activities according to each schedule of the resource to be used (business days, holidays, etc.)? | 7 |
| 10 | Is there a process for identifying and prioritizing the critical path of the project? | 7 |
| 11 | Is there a baseline recording of the project at the end of the schedule development? | 7 |
| 12 | In developing the schedule, do all activities have resources? | 7 |
| 13 | Is there a formalization process of generating provisions (contingencies) for project costs? | 7 |
| 14 | Is the process of formalizing a matrix of responsibilities for the project? | 0 |
| 15 | Is there a formalized process of generating and transmitting project information during its execution (Communications Plan)? | 0 |
| 16 | Is there a mapping that details who needs the information, when it will be needed, how it should be defined, and by whom? | 0 |
| 17 | Is there a place with phones, addresses, e-mails and contacts from the entire team and involved? | 7 |
| 18 | Is there a formalized process for defining the schedule of project meetings? | 0 |
| 20 | Are the meetings pre-scheduled from the start and on the agenda of the entire team? | 0 |
| 21 | Does each team member know what information to bring to the follow-up meeting? | 7 |
| 22 | Is there a formal project budgeting process by grouping estimates of activities, resources, hiring, contingencies, etc.? | 7 |
| 23 | Is there a formalized risk identification process for the project? | 0 |
| 24 | Is there a qualitative and quantitative project risk assessment process? | 0 |
| 25 | Is there a project risk prioritization process? | 0 |
| 26 | Is there a process for planning the project's risk responses? | 0 |
| 27 | Is there a process to identify who is responsible for the project risk? | 0 |
| 28 | Is there a risk management and monitoring tool for the project? | 0 |
| 29 | Is there a process for identifying project product requirements? | 7 |
| 30 | Are the project product requirements translated into the work required to develop this product? | 7 |
| 31 | Is there a document that can formalize the work of the project (Scope Statement)? | 7 |
| 32 | Is there a formalized audit plan for quality requirements for the project (Quality Plan)? | 7 |
| 33 | Is there formalized planning for the provision and contracting of project resources (engineering, equipment, etc.)? | 7 |
| 34 | Is there a formalized contracting cycle for large project deliverables (Bid Cycle or Hiring Cycle)? | 7 |

*Table 92. Performance and results*

| Id. | Question | Score |
|---|---|---|
| **01** | Are deadlines met as planned? | 7 |
| **02** | Are the quality of the delivered projects according to the customer's requirements? | 7 |
| **03** | Are project costs met as planned? | 7 |
| **04** | Is there a monitoring of customer satisfaction with the project product? | 7 |
| **05** | Is there a process for evaluating the performance of the project manager in the conduct of the project? | 7 |
| **06** | Is there a project team evaluation process? | 0 |

*Table 93. Methods and practices*

| Id. | Question | Score |
|---|---|---|
| **01** | Before the start of the project is a feasibility study for the investment carried out? | 7 |
| **02** | Is there a practice of benchmarking projects in relation to projects developed by similar companies? | 7 |
| **03** | Is there practice of contracting consultancies to identify and evaluate project maturity for successful projects (IPA - Independent Project Analysis, etc.)? | 7 |
| **04** | Is there practice of hiring consultancies in the evaluation and optimization of budget? | 0 |
| **05** | Is there a formal process of closing contracts open during the project? | 7 |
| **06** | Is there a formal process of collecting the lessons learned during and at the end of the project? | 7 |
| **07** | Is there a formal process of cost control through tools like EVA - Value Added Estimation? | 7 |
| **08** | Is there a formal operational risk analysis and survey process (HAZOP - Hazard and Operability Studies)? | 0 |

# Table 94. Engineering and geology

| Id. | Question | Score |
|---|---|---|
| 01 | Is there a formalized process that every project developed in the organization should go through the phases: conceptual, basic and detailed? | 0 |
| 02 | Is every project being run by the organization going through or going through the conceptual, basic, and detailed engineering phases? | 0 |
| 03 | Is there a formalized process for the drilling and auditing of mineral resources for engineering? | 7 |
| 04 | Is there a formal process that all conceptual engineering should begin with the first results of exploratory surveys? | 7 |
| 05 | Is there a formalization and submission of conceptual engineering information to the Environmental area for environmental licensing? | 7 |
| 06 | Is all Conceptual Engineering developed based on a mapping of biotopes (areas impacted by the environment) and surface? | 7 |
| 07 | Are all process tests (Technological Development) formally sent to Engineering to start Basic Engineering? | 7 |
| 08 | Is there a formalized process for reviewing engineering quality (Peer Review) before moving from one phase to another (e.g., Conceptual Engineering for Basic Engineering)? | 7 |
| 09 | Is there a formalized engineering development culture in 3D technology? | 0 |
| 10 | Is there a process of evaluation, prior to the beginning of the Detailed Engineering, on the detailing of the project through the modularization of the construction? (Construction in Modules) | 0 |
| 11 | Is there the process of generating a Master Plan for the project after Conceptual Engineering? | 0 |
| 12 | If yes for answer 11, is there a formalized process of freezing this master plan and communicating it to all project stakeholders? | 0 |
| 13 | Is there a formalized process of generation of Descriptive Memorials, Spreadsheets, Data Sheet and Technical Specifications for the Supplies area? | 7 |
| 14 | Is there a formalized process of participation of the Supplies area during the evolution of the Basic Engineering for Detailed Engineering to determine the type of contracting of the Supply, Assembly and Implementation of the project? | 7 |
| 15 | Is there a formal process of realizing and studying tradeoffs for the lowest cost for utilities (water, gas, electricity, logistics, etc.)? | 7 |
| 16 | Is there a formal process of realization and study of opportunities (tradeoffs) regarding the location of the project (surface workers, communities, logistics, etc.)? | 7 |
| 17 | Is there a formal process for evaluating and validating each engineering phase (Conceptual, Basic, and Detailed) before moving from one to the other (Front End Loading Portions of Independent Project Analysis)? | 0 |
| 18 | At the end of the Project's Conceptual Engineering, is there a formal process to generate a Prefeasibility Study? | 7 |
| 19 | At the end of the Basic Engineering of the Project, is there a formal process to generate a Feasibility Study of the project? | 7 |
| 20 | During the execution of the Conceptual Engineering, is there a formal process of identification and first contacts with impacted landowners as well as local communities? | 7 |
| 21 | During the execution of Conceptual Engineering, is there a formal topographic survey of the region as well as the generation of orthophotos for the project's master plan? | 7 |
| 22 | After Conceptual Engineering, is there a formal process of generating investment estimates at a precision level of -15% to + 25%? | 7 |
| 23 | After the Basic Engineering, is there a formal process of generating the investment estimates at a precision level of -10% to + 10%? | 7 |
| 24 | Is there a formal closed-end planning process for the project after its useful life? | 7 |
| 25 | Is there a formal process for identifying VIPs - Value Improving Practices? | 0 |

*Table 95. Organizational structure*

| Id. | Question | Score |
|---|---|---|
| **01** | When a project starts, is there a project manager assigned by the organization to lead the project? | 7 |
| **02** | Is there an area focused on project planning? | 0 |
| **03** | Is there a formalized position of project manager within the organization? | 0 |
| **04** | Is there formalization of the Project Manager's power to similar functional positions (HR Manager, Operations Manager, etc.)? | 7 |
| **05** | Is there formalization of areas within Project management such as: Planning, PMO, Cost Control, etc.? | 0 |
| **06** | Are all internal areas of the company formally communicated from the beginning of the project and are their representatives formally assigned to the project? | 7 |

*Table 96. People management*

| Id. | Question | Score |
|---|---|---|
| **01** | Is there a formal process of developing the project manager's career and team? | 0 |
| **02** | Do follow-up meetings have a schedule and a fixed format? | 0 |
| **03** | Does each project member receive their work with instructions and in advance? | 0 |
| **04** | Is there a formal process of variable compensation for project team performance? | 0 |
| **05** | Is there a formal process for evaluating project team performance? | 0 |
| **06** | Is there a formal organization chart of the project team, delegations, etc.? | 7 |
| **07** | Are all the human resources required for the project formally sent to the HR area to arrange recruitment? | 7 |
| **08** | Are all interpersonal skills of the project team managed during the project? | 7 |
| **09** | Are all conflicts between the project team formally addressed? | 0 |

## Table 97. Project management office

| Id. | Question | Score |
|---|---|---|
| 01 | Is there a flow for the arrival in your area of the information about the existence of a project or a contract? | 7 |
| 02 | Is there a dedication of the professionals to the management of the projects? | 0 |
| 03 | Is there any tool for estimating effort (time) and project costs? | 7 |
| 04 | Is there any tool for estimating project resources? | 7 |
| 05 | Is there any tool to formalize the client's acceptance? | 7 |
| 06 | Is there an established process for formal acceptance of each deliverable of the project? | 7 |
| 07 | For the development of project documentation, is there a repository of past project templates for reuse in new projects? | 7 |
| 08 | Is there an area focused on the physical and financial control of the project? | 7 |
| 09 | Is there a visibility room, a mural or even an intranet site where everyone can access projects in a simple way? | 7 |
| 10 | In the case of geographically dispersed teams, are the procedures / technologies used to integrate them adequate? | 7 |
| 11 | Are there managerial reports with a previously defined format agreed upon with the client? | 7 |
| 12 | Is there a formalized process for the closing of a project / contract? | 7 |
| 13 | Is there a formal process for the provision of information in the closing phase (finalization of the work)? | 7 |
| 14 | There is an area focused on the control and support of the project (PMO or Project Office) | 0 |

## Table 98. Stakeholder management

| Id. | Question | Score |
|---|---|---|
| 01 | Is there a formalized process of mapping and managing the company's internal stakeholders? | 0 |
| 02 | Is there a formalized process for managing stakeholder expectations of the project during its execution? | 0 |
| 03 | Are formal interviews conducted to identify project stakeholders? | 0 |
| 04 | Are all potential impacts caused by each stakeholder formally assessed and managed during and after the project? | 0 |
| 05 | Is there a formal process of ranking and prioritizing the degree of influence of each stakeholder on the project? | 0 |
| 06 | Is there a formalized process of classification of stakeholder groups? | 7 |
| 07 | Is there a formal process of simulating the reactions of each stakeholder to the various situations that may occur during and after the project? | 7 |
| 08 | Is there a formalized process of gathering and filing the information of each stakeholder such as: name, address, telephone, etc.? | 7 |
| 09 | Is there a formalized Communication Plan capable of identifying what kind of information, when, how and for which stakeholder to send? | 0 |

*Table 99. Environment management*

| Id. | Question | Score |
|---|---|---|
| 01 | Is there a formalized process of beginning environmental studies with the beginning of Conceptual Engineering? | 7 |
| 02 | Is there a formalized process for mapping biotopes (areas of environmental impact) to aid in the development of Conceptual Engineering? | 7 |
| 03 | Is there a formalization process of mapping the surface properties impacted by the project during the Environmental Licensing and Conceptual Engineering? | 7 |
| 04 | Is there a formal process of generating documentation for Environmental Impact Studies and Environmental Impact Report by the Engineering area? | 7 |
| 05 | Is there a formal process that synchronizes the development of Engineering and Environment activities? (Generation of Documentation for Environmental Licensing)? | 7 |
| 06 | Is there a formalized area for institutional relations with government agencies? | 7 |
| 07 | Is there a formalized area for relationship with the communities impacted by the project? | 7 |
| 08 | Is there a formalized integration of the internal areas of the company: Engineering, Environment, Institutional Relationship and Social Responsibility? | 7 |
| 09 | Is there a formal process to monitor the release of Pre-Installation, Installation and Operation licenses from government agencies? | 7 |

# About the Authors

**João Carlos Araújo da Silva Neto** is Head of CAPEX and Plant Technology in RHIMagnesita and graduating in Innovation in Stanford University. Also a PMP Certificate, APMC Certificate, Master in Strategic Administration, Electrical Engineer and Post Graduated in Project Management, Business Management, Portfolio Management. Experienced project manager for projects in sectors as mining, oil and gas, civil construction and steel.

**Ítalo Coutinho** is Engineering Manager of Saletto Engenharia, PMP Certificate, Doctoring in Business Administration, Mechanical Engineer, Post-Graduate in Project Management. Author and presenter about Industrial Project Management. PMI volunteer for the PMBOK Construction Extension. Experienced construction manager for projects in sectors as mining, oil and gas, civil construction, malls and steel.

**Gustavo Teixeira** is Managing Director of PMBASIS, PMP,PROSCI and PMO-CP CertificatE. Master in knowledge management, System analyst, Post-Graduate in Business management. Author and presenter about Project Management. PMI volunteer. Experienced technology manager for projects in sectors as health, services, telecom and consultant.

**Alexandro Avila de Moura** is Chief Operating Officer in Paranapanema. Degree in Mechanic Engineering by Universidade Federal do Rio Grande do Sul (UFRGS). Post-graduation in Strategic Management of Business by Universidade de São Paulo (USP). Post- graduation in Project Management by Faculdade Pitágoras. Master in Administration (incomplete) by Pontifícia Universidade Católica – MG. More than 25 years working in Brazilian Mining and Metals Sector, in charge of operations and projects development. Experience in occupational health and safety, environment and community relationship also. Certified as PMP (Project Management Professional) by PMI (Project Management Institute). Author of blog "Gestão de Produção com Alexandro" [Production management with Alexandro] in Brazilian Portuguese (www.gestaoproducaocomalexandro.com).

# Index

## F



## H



## I



## M



## O



## P



## R

## S



## T



## V

www.ingramcontent.com/pod-product-compliance
Lightning Source LLC
LaVergne TN
LVHW061539290926
849162LV00004B/842